SpringerBriefs in Astronomy

Series Editors

Martin Ratcliffe
Michael Inglis
Wolfgang Hillebrandt

For further volumes:
http://www.springer.com/series/10090

Charles J. Byrne

The Moon's Near Side Megabasin and Far Side Bulge

 Springer

Charles J. Byrne
Middletown
NJ
USA

ISSN 2191-9100   ISSN 2191-9119   (electronic)
ISBN 978-1-4614-6948-3 ISBN 978-1-4614-6949-0   (eBook)
DOI 10.1007/978-1-4614-6949-0
Springer New York Heidelberg Dordrecht London

Library of Congress Control Number: 2013933587

© Charles J. Byrne 2013
This work is subject to copyright. All rights are reserved by the Publisher, whether the whole or part
of the material is concerned, specifically the rights of translation, reprinting, reuse of illustrations,
recitation, broadcasting, reproduction on microfilms or in any other physical way, and transmission or
information storage and retrieval, electronic adaptation, computer software, or by similar or dissimilar
methodology now known or hereafter developed. Exempted from this legal reservation are brief excerpts
in connection with reviews or scholarly analysis or material supplied specifically for the purpose of
being entered and executed on a computer system, for exclusive use by the purchaser of the work.
Duplication of this publication or parts thereof is permitted only under the provisions of the Copyright
Law of the Publisher's location, in its current version, and permission for use must always be obtained
from Springer. Permissions for use may be obtained through RightsLink at the Copyright Clearance
Center. Violations are liable to prosecution under the respective Copyright Law.
The use of general descriptive names, registered names, trademarks, service marks, etc. in this publication
does not imply, even in the absence of a specific statement, that such names are exempt from the relevant
protective laws and regulations and therefore free for general use.
While the advice and information in this book are believed to be true and accurate at the date of
publication, neither the authors nor the editors nor the publisher can accept any legal responsibility for
any errors or omissions that may be made. The publisher makes no warranty, express or implied, with
respect to the material contained herein.

Printed on acid-free paper

Springer is part of Springer Science+Business Media (www.springer.com)

Preface

In our human history, we have perceived the Moon in different ways. Early astronomers saw the Moon as a sphere, the only object in the sky except the sun that was clearly not a point. Different cultures identified the patterns of light and dark in diverse ways, such as a man in the Moon and a rabbit. Now, we call the dark areas maria, comparing them to Earth's seas. Shortly after the invention of telescopes, Galileo Galilei and others turned them to the Moon and saw that it was not a perfect sphere but was heavily cratered.

With or without instruments, we here on Earth see only the near side of the Moon: the far side is known to us only by the few astronauts who have orbited around it and the many unmanned space missions that have been sent around it. This is typical of a moon of a large planet because the gravity field of the planet dissipates any initial relative rotational moment of the moon by inducing a tidal effect. As a result, the Moon is locked so that its rotational period is equal to its 28 days period of revolution around Earth.

As spacecraft viewed the far side of the Moon, we learned that it looks very different from the near side; there is almost no pattern of dark smooth areas against lighter heavily cratered highlands. Later, manned and unmanned missions (for example, Apollo, Clementine, Chang-E, Kaguya, and Lunar Reconnaissance Orbiter) carried altimeters that defined the surface shape of the Moon.

Now we have Digital Elevation Maps (DEMs) for the topography of the whole Moon, defining its surface shape precisely. Most of the far side is covered by a bulge rising 6 km high, while the Man in the Moon area on the near side is about 2 km low and mostly level. The surface topography is only part of the description of the shape of the Moon. Tracking of spacecraft has produced a progressively refined measurement of the gravity field of the Moon that, in combination with topography, provides extensive information about the subsurface density structure, including the shape of the boundary between crust and mantle. The thickness of the crust is up to 26 km thicker on the far side than on the near side.

With the detailed DEMs, we can model the larger impact features one by one, and build a library of such features. As we near completion of the set of models, a comparison with the DEM of the measured topography reveals omissions and imperfections in the library of models, revealing new impact features and a few features that have other causes. The set of modeled features described here reduces the differences between the comprehensive model and the measured topography to such an extent that the residual error is free of obvious features larger than about 200 km, a small fraction of the 8,696 km circumference of the Moon.

Samples returned to Earth by Apollo missions, supplemented by remote sensing instruments of increasing sophistication, give us insights into variations in composition and ages of some of the features, inferred from the ages of the soil and rocks exposed on the surface. Detailed descriptions of such mineralogy are not presented here but the results of analysis are used to support the sequence and age estimates of modeled features.

In this book, an algorithmic model of hypervelocity impact features is developed, based on extensions to the Maxwell-Z model for the ejection of material from an apparent crater. An extension permitting extrapolation of the model to impact features of arbitrarily large sizes is presented. A single algorithmic model serves for all impact features with the primary parameters being center coordinates, apparent diameter, apparent depth, and depth of level fill (if present).

In 2007, I found a major feature that is needed to explain the first-order asymmetry between the lunar near side and far side. I called it the Near Side Megabasin. This feature, similar to familiar impact basins but much larger, explains the topography and crustal thickness of the near side and (with its ejecta) the bulge on the far side. There is a long history of related proposals by J. Wood, P. Cadogan, E. Whittaker and others, but no previous explanation has stood up to quantitative comparison between theory and observations. The Near Side Megabasin was proposed on the basis of Clementine data and it fits new data from later missions, as well as new understanding of large impact basins, even better.

With the new data, the model of the Near Side Megabasin allowed derivation of an improved model of the South Pole-Aitken Basin (also an early megabasin) and the Chaplygin-Mande'shtam Basin, another previously unknown early megabasin. Next, a topographic model composed of these giant features and models of other basins and large features was constructed. Comparison of the composite model with the observed topography resulted in the discovery of several new features first reported here.

The set of modeled features are placed in time sequence by geologic period to produce a history of the Moon from when its crust first solidified until today. The larger pattern of this history follows the traditional stratigraphic periods established by Gene Shoemaker and modified by Don Wilhelms.

The lunar history in this book incorporates new observations and new understanding of giant impact features, of the era when the planets were forming. The early megabasins are proposed to have been caused by some of the last planetesimals that were completing the accretion of Earth. For each such event on the Moon, there were about 17 such impacts on Earth during the Hadean period, before the origin of life.

The lunar Late Heavy Bombardment (LHB), a period of a high rate of lunar impact events in the Nectarian, and Lower Imbrian periods, was proposed by F. Tera and others in 1974, based on the concentration of ages of Apollo samples. The lunar LHB is now understood to be directly related to a new model of the evolution of the outer solar system, developed by astronomers at Nice, France and subsequently extended to the inner solar system. In the history of Earth, the time of the LHB occurred during the Early Archean period, shortly before or while primitive life was evolving. The extended Nice Model also identifies the probable source of most of the impacting asteroids after the LHB and the subsequent decline in size and rate of impacts.

Acknowledgments

To the extent that this book contains statements on lunar geology and geologic history, I am indebted to Don E. Wilhelms for his patient instruction and corrections in the past with two previous Springer books as well as for this book. I, of course, am responsible for any remaining problems in those fields. Don has been extraordinarily patient and supporting, doing his best with a student of limited talent.

In addition, I have benefited from encouragement and guidance from several people who have encouraged me along the way. These include Paul Lucie, Paul Spudis, Greg Neumann, Ray Hawke, and Bill Bottke. The instigation for the book is the responsibility of Maury Solomom, publishing editor of Springer, who suggested this SpringerBrief.

My much-loved and loving wife Mary has endured my distraction during parts of the preparation of this book, as well as reviewing and rereviewing the work as it has progressed.

Contents

Chapter 1
The Dichotic Shape of the Moon: Differences Between the Lunar Near Side and the Far Side

1.1 Introduction

When the Moon is viewed in its entirety, the most striking difference is the contrast between the near side and far side: this dichotomy extends to brightness, elevation, crater density, mineral composition, gravity anomalies, thickness variations of the crust, and the center of gravity offset relative to the center of figure. In the following sections, each of these qualities is examined, based on the voluminous data currently available from instruments carried by spacecraft in lunar orbit. In this chapter, these differences are discussed qualitatively. In later chapters, quantitative models explaining these differences are proposed.

1.2 Brightness Patterns

The Clementine spacecraft was placed in a polar lunar orbit aligned with the sun (zero phase angle) in order to receive relatively uniform brightness for spectroscopy. As a result, the images recorded albedo, the inherent reflectivity of the surface, especially near equatorial areas. Since the camera pointed down toward the surface, the phase angle changed with latitude, but differences in albedo were still recorded. A comparison of the resulting images of the near side and far side of the Moon is shown in Fig. 1.1.

The very dark area on the near side in Fig. 1.1 is mostly basaltic mare material. The somewhat lighter gray area in the southern hemisphere of the far side is in the crater of the giant South Pole-Aitken Basin. While there are some basalt mare flows in the far side, most of the surface there is bright highland areas that only rarely appear on the near side.

C. J. Byrne, *The Moon's Near Side Megabasin and Far Side Bulge*,
SpringerBriefs in Astronomy, DOI: 10.1007/978-1-4614-6949-0_1,
© Charles J. Byrne 2013

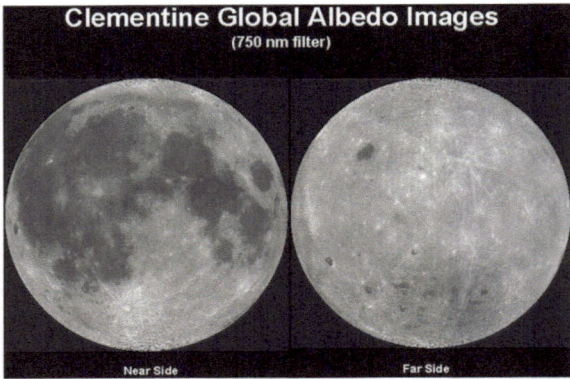

Fig. 1.1 Brightness images for the near side (*left*) and far side (*right*) of the Moon from Clementine data (Lambert equal area projections). Note that *the dark maria* cover much more of the near side than the far side. *Source* NRL, Clementine, LPI, Paul Spudis: http://www.lpi.usra. edu/lunar/missions/clementine/images/

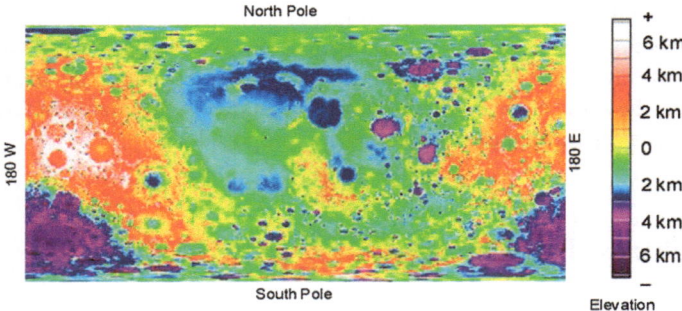

Fig. 1.2 This is a Digital Elevation Map (DEM) from Kaguya data. *The false color scale* for elevation is by this author. This is a geometric projection, a graph of elevation versus latitude and longitude. *Each pixel* represents the elevation of an area 1° by 1°. The near side is centered and the far side is split to the east and west. Horizontal distances are progressively stretched at northern and southern latitudes. Note that elevations on the near side are mostly limited to the range between 0 and −2,500 m, while the far side elevations are in the range of −7,000 m (the floor of the South Pole-Aitken Basin) and +6,000 m (the far side bulge). *Source* JAXA, Kaguya: http://www.kaguya.jaxa.jp/ en/science/LALT/The_lunar_topographic_data_e.htm, File name: LALT_G.txt

1.3 Elevation

The near side has a relatively level circular area sloping up to a bulge on the far side (Fig. 1.2). The South Pole-Aitken Basin has impacted this bulge, forming the deepest area on the Moon.

The elevation data is the clearest indicator of the dichotic nature of the asymmetry between the near side and the far side. Further, it is the tool for modeling the largest features on the Moon. In addition to the DEM of Fig. 1.2, the 1/16° by 1/16° DEM of Kaguya is used to model the smaller features.

1.4   Crater Density

In the course of the Moon's history, parts of its landscape were resurfaced, erasing evidence of past bombardment. Such resurfacing included basin formation (encompassing the ejecta blanket from basins and large craters, typically out to twice the radius of the feature), volcanic lava flooding (not only in named maria but also in sheets connecting them), deposition of smooth plains (beyond the ejecta blanket), and by pyroclastic eruption (typically beyond the edges of deep maria).

The areas of resurfacing can be detected by the density of relatively small craters. As the history of the Moon progressed, the intensity of impacts dropped precipitously after the period of heavy bombardment. The resolution of Fig. 1.2 is sufficient to distinguish areas that were resurfaced about 3 Gigaannum (Ga) years ago from older areas. Nearly all of the resurfaced area is on the near side (mostly north of 40°S and east of 40°E), indicating strong volcanism there after the last period of heavy bombardment.

1.5   Mineral Composition

Remote sensing instruments on several spacecraft missions were able to detect anomalies in element and mineral concentrations. Examples of element anomaly data are shown in Figs. 1.3 and 1.4.

The minerals showing similar anomalies included iron, titanium, thorium, and the associated group of incompatible elements known as potassium, rare earth elements, and phosphorus (KREEP). Nearly all the significant concentrations of these elements are found in areas of the near side. Samples collected by the Apollo astronauts (from widely varied terrain but all from the near side) have provided ground evidence to confirm the remote sensing data. These element anomaly maps are in close agreement. The major anomalies are on the near side in the region

Fig. 1.3 Iron anomaly map. NRL, Clementine, LPI, Paul Spudis

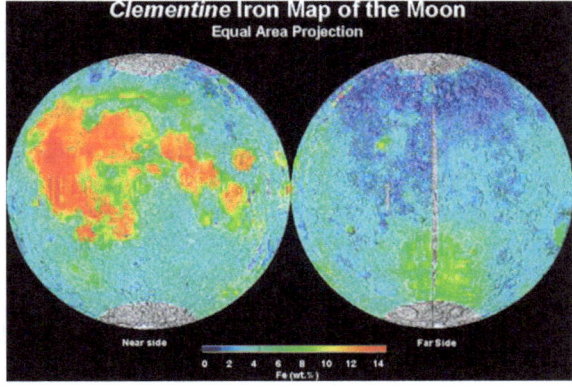

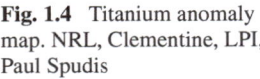

Fig. 1.4 Titanium anomaly map. NRL, Clementine, LPI, Paul Spudis

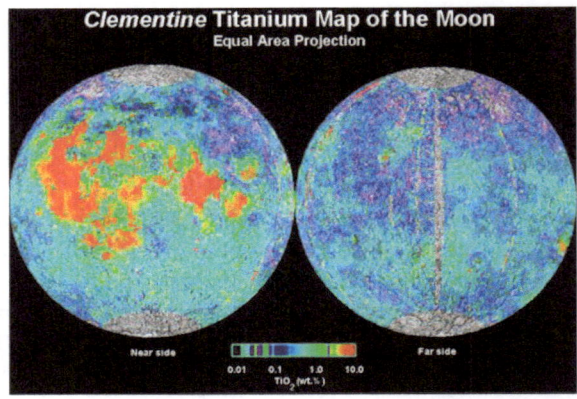

where there has been extensive resurfacing. Mild anomalies occur within the South Pole-Aitken Basin, where the cavity is deep.

1.6 Gravity Anomalies and Thickness of the Crust

It would be nice to have drill cores to the order of 60 km of depth. Unfortunately, the state of the art for core samples on the Moon was more like 60 cm in the days of Apollo. Another way to measure the depth of the boundary between the crust and mantle is through monitoring seismic disturbances, since the velocity of sound varies between the mantle and crust. However, such data is available only on the near side, where seismometers were deployed at Apollo landing sites.

Through remote sensing, global crustal thickness (see Fig. 1.5) is estimated by relating topography to free-air gravity anomalies. A mountain or depression and/ or a variation in the density of the surface material may affect the free-air gravity value. If the gravity value is corrected for the topography, it is called Bouguer gravity (Zuber et al. 1994). If an anomaly in free-air gravity is not present, then any extra mass or reduction in mass implied by topography variations must have been compensated for by a complementary movement of the underlying dense mantle material. This is called isostatic compensation. A map of the inferred crustal thickness variations is shown in Fig. 1.5.

The crustal thickness under the far side bulge is seen to be very deep in the vicinity of the Korolev Basin, and is deepest at the intersecting rims of the Korolev and Dirichlet-Jackson basins. The near side crust is nearly uniformly thin in the resurfaced area and just slightly thicker in the additional area of level depression extending near the eastern limb (see Fig. 1.2). A thinner arc extends from the Nubium Basin through Oceanus Procellarum and Mare Frigoris. The crust is also thin beneath the deep floor of the South Pole-Aitken Basin. A further discussion of gravity measurements is in Chap. 7.

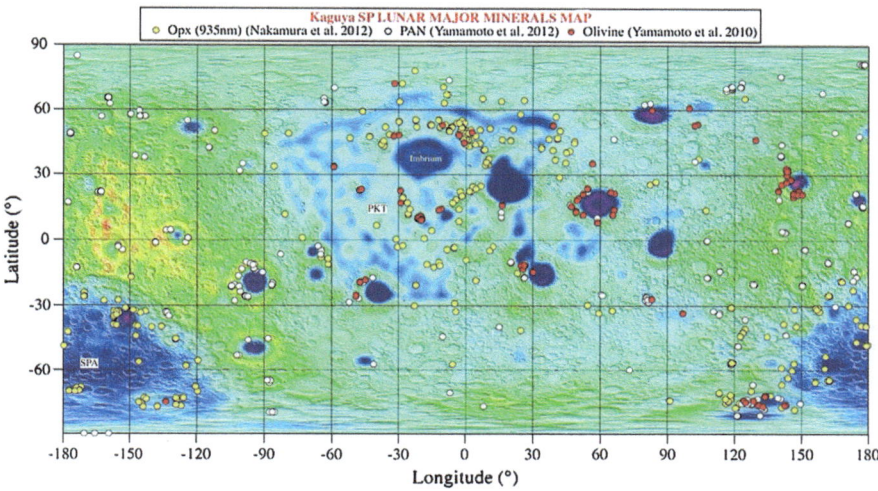

Fig. 1.5 Base map, crustal thickness derived from Kaguya gravity and topography data (Nakamura et al. 2013), *yellow* is high and purple is low. Small circles show outcrops of crystaline rocks detected by Kaguaya spectroscopy. *White* is pure anorthosyte (PAN), *yellow* is low calcium pyroxine (LCP), and *red* is olivine

1.7 Crystalline Outcrops in Impact Craters

Recent high-resolution spectroscopy has revealed small outcrops of crystallized minerals in material exposed in central peaks and rims of craters. Unlike most of the glassy lunar regolith, some of these outcrops have retained their original structure and display clear spectroscopic patterns. Nearly all of these outcrops are on the far side highlands (see Fig. 1.5).

1.8 Offset of the Center of Gravity

The center of mass of the Moon has been calculated to be 2.5 ± 0.4 km closer to the near side than the center of figure (Haines and Metzger 1980). This offset is consistent with the crustal thickness calculations. The thicker, lower density crust on the far side extends the center of figure back from the near side, equivalent to shifting the high density core and mantle forward.

Chapter 2
Hypervelocity Impact Features from Craters to Ringed Basins to Megabasins

2.1 Introduction

The topography of the Moon is dominated by the cavities and ejecta caused by hypervelocity impacts. Over the course of the history of the solar system, all rocky bodies have, in the course of their development, been subjected to bombardment by collisions with other bodies of both rocky and icy composition. The intensity of this bombardment has waxed and waned. Fortunately for us it has been decreasing to a small but still hazardous rate. Each collision is chaotic, yet the statistical properties of ensembles of events are regular and subject to quantitative analysis and modeling (Croft 1980; Housen et al. 1983; Byrne 2007).

The observable shape of the result of each event, its morphology, is quite clear on rocky bodies that lack a substantial atmosphere such as the Moon, Mercury, and the asteroids. They are said to be a "witness plate" of the history of the solar system, often showing the sequence of events, but not the absolute age of the events. Absolute age is only obtained through analysis of radioactive isotopes of samples either collected in situ or, fortuitously, meteorites collected on Earth that can be assigned to specific bodies or sites by analysis of their mineral composition.

A quantitative model is established in the next chapter. This discussion is limited to the dominant topographic features of impact craters and their rims and ejecta fields, especially those that change with the size of the impact event. The qualitative nature of impact events is discussed in this chapter.

2.2 Nomenclature

The terminology for impact features has changed as understanding has advanced. The term recognized by the International Astronomical Union (IAU) is crater. It is defined as a bowl-shaped depression, independent of its cause, and is given a name of a person who has been involved in its description or is being honored. Therefore, a crater

C. J. Byrne, *The Moon's Near Side Megabasin and Far Side Bulge*,
SpringerBriefs in Astronomy, DOI: 10.1007/978-1-4614-6949-0_2,
© Charles J. Byrne 2013

in the IAU list can be caused by impact, volcanism, or subsidence. This rule allows assigning a distinguishing name to a feature while its origin is still a matter of investigation. The term "basin" was introduced by the United States Geological Service (USGS) to describe circular features that enclosed maria or had other characteristics suggesting similarities. If the surrounded mare had an IAU name (such as Mare Imbrium), or if the cavity itself had been named as a crater (Bailly), those names were adopted for the basin by the USGS (Wilhelms 1987). Otherwise, names were assigned by combining two features, usually named craters, superposed on the basin's crater (Coulomb-Sarton Basin, South Pole-Aitken Basin) (Wilhelms 1987). Craters were considered basins if the diameter assigned in the literature was greater than 300 km. In this work, a basin is considered to be a type of crater. In retrospect, it may have been better to call all of these features craters, give them single formal names, and then proceed to describe their discernable properties. At this time however, the term basin is too well established in common usage for a change to be worthwhile.

Nearly all craters on the Moon are either primary craters caused by hypersonic impacts from beyond the Moon or secondary craters caused by boulders ejected from primary craters. A hypersonic impactor arrives at a velocity that exceeds the speed of sound in the target material. If the approach velocity is approximately vertical (within about 60°) to the target surface, the feature will be approximately circular and the result is similar to setting off a sufficiently intense explosion that generates a shock wave in the target surface. The target material is compressed beyond its normal range of elasticity as energy is transferred from the shock wave. Eventually, the velocity of the shock wave drops to the sonic velocity in the target material, which thereafter transmits the remaining energy with minimal attenuation. After the shock wave passes a particular location, the material there reacts depending on its composition and local environment. The energy imparted by the shock wave is released in various forms: kinetic energy, fracturing, rearrangement of crystal structure, elastic compression, heating, melting, and vaporization.

Part of the historical difficulty in naming impact features is the gradual change in morphology with size: also, it has been difficult to arrive at a common set of terms for both terrestrial and lunar craters, partly because of the difference in erosion phenomena. Recently, a set of widely accepted terms has been adopted to describe hypersonic impact craters for all rocky bodies (Turtle et al. 2004).

Hypersonic impact features of all sizes have a central cavity, a rim raised above the original floor (the pre-impact target surface), and an ejecta field formed from material thrown out of the cavity. The following terms have gained wide acceptance:

- Transient cavity: The region bounded by the transition from the expanding shock wave to the sonic wave. Some of the target material within this boundary is ejected and some of the remainder is transformed by the shock wave.
- Apparent crater: The crater after redistribution of material left in the transient cavity plus material contributed by early landslides from the walls of the transient cavity.
- Apparent crater diameter: The diameter of the apparent crater measured at the intersection with the target surface.
- Peak: Some apparent craters have central peaks.

- Peak-ring: Some apparent craters have elevated rings within them.
- Inner ring diameter: The diameter of the circle that best fits the highest points of peak-rings or other rings within the apparent crater.

2.3 Classification of Impact Crater Morphology

Impact craters come in many different forms depending on their size and the nature of the pre-impact target surface. The following description of sub-classes is widely, but not universally, used. The size ranges are only approximations; the diameters of craters in the different subclasses overlap. Many impact features fall neatly into these categories, but some have transition shapes that fall between categories.

2.3.1 Simple Craters

Simple craters, those typically smaller than about 10 km, usually have bowl-shaped cavities. Craters in the approximate range of 10–100 km usually have a central peak. An example of the simple crater Linné is shown in Fig. 2.1 and a typical cross-section is shown in Fig. 2.2. Linné was formed in basalt and has experienced little modification since its formation. The depth to diameter ratio of simple craters, about 0.2, is independent of size.

2.3.2 Complex Craters

Craters in the range of about 10–150 km usually have more detailed structure than simple craters. The smaller craters in this range have terraces and cliffs just inside their walls. Larger complex craters have central peaks or clusters of peaks composed of rugged crustal material and are thought to result from elastic rebound of the material below the transient cavity. As the crater size of complex craters increases, the

Fig. 2.1 Linné, a 2.22 km simple crater in northwestern Mare Serenitatis (27.7°N and 11.8°E). This image combines a Digital Elevation Map (DEM), 60 m resolution with a camera image, 2 m resolution. *Source* NASA, Lunar Reconnaissance Orbiter, (Garvin et al. 2011)

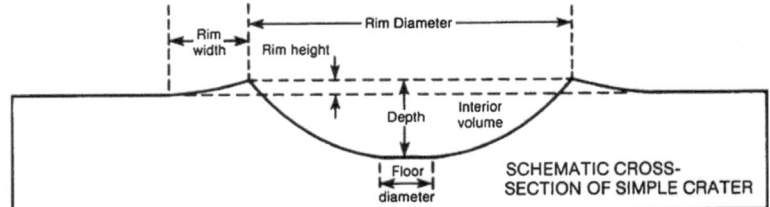

Fig. 2.2 Cross-section of a simple crater. The parameters here are as often measured in the literature but apparent diameter and depth are measured from the target surface. Rim diameter is sometimes called rim crest diameter. *Source* Lunar Sourcebook, Fig. 4.2 (Heiken et al. 1991)

Fig. 2.3 Tycho (apparent diameter, 74 km) is a complex crater with a central peak, located in the central near side highlands (43.3°S and 11.3°E). *Source* NASA, Lunar Orbiter, LPI, (Byrne 2008)

central peak is replaced by a roughly circular region of hills called the peak-ring. Tycho, an example of a complex crater with terraces and a central peak, is shown in Fig. 2.3 and its radial elevation profile is shown in Fig. 2.4. Complex craters (and the larger types as well) become progressively more shallow than simple craters.

2.3.3 Peak-Ring Craters

As the craters grow larger than 150 km, the central uplift structure takes the form of a circular ring of peaks. Such features are called peak-ring craters. The Schrödinger crater (also called a basin because of its size) is shown in Fig. 2.5 and its radial profile is shown in Fig. 2.6. The well-formed inner ring of peaks, like most single inner rings, has a diameter of about half the apparent diameter. This peak-ring is nearly continuous.

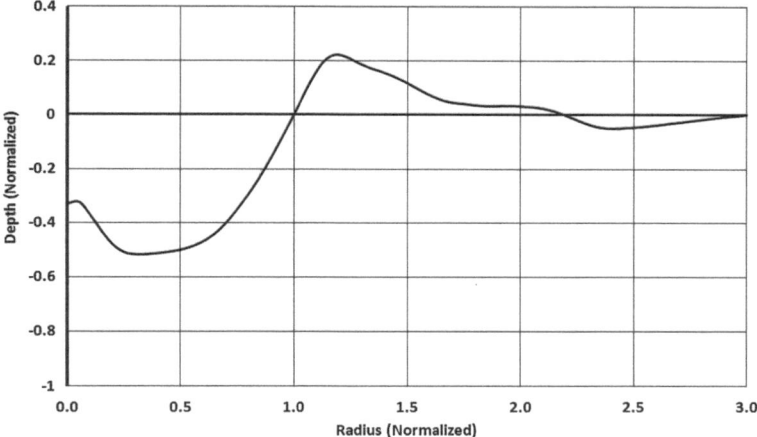

Fig. 2.4 A radial elevation profile of Tycho (elevation averaged over 360° in azimuth). *Source* JAXA, Kaguya, 1/16° DEM, current author

Fig. 2.5 The Schrödinger Basin (apparent diameter is 298 km) is a peak-ring crater. It is located at 79.8°S, 133.2°E in ejecta from the South Pole-Aitken Basin. *Source* NRL, Clementine, http://en.wikipedia.org/wiki/File:Schrodinger_crater.gif

2.3.4 Multi-Ring Basins

Craters that are larger than 300 km are usually called basins. They may develop additional rings, inside and/or outside of their rims; such features are called multi-ring basins. Note that a peak-ring crater has 2 rings, the peak-ring and the main ring at the wall of the transient cavity, but at least one additional ring is needed to be considered "multi-ring". The Orientale Basin is the archetype of a multi-ring basin (Fig. 2.7). It has three inner rings and arcs of two outer rings. The eastern part of the Orientale Basin descends to the east, down the slope of the cavity of

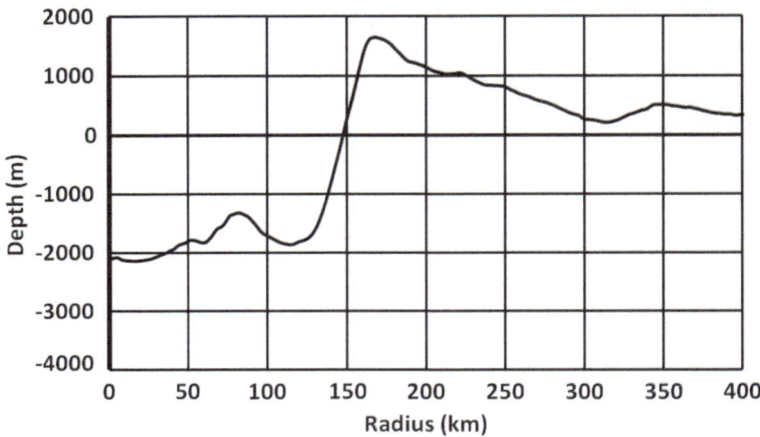

Fig. 2.6 Radial profile of the Schrödinger Basin (apparent diameter, 298 km). *Source* JAXA, Kaguya, 1/16° DEM, current author

Fig. 2.7 The Orientale Basin (apparent diameter, 890 km) is the youngest example of a multi-ring crater. It is located at 19.6°S and 95.0°W on the rim of the Near Side Megabasin. This is a topographic map of the basin derived from Lunar Reconnaissance Orbiter data. *Source* NASA, GSFC, MIT, Brown U., GSA

the Near Side Megabasin (NSM). The central and western part rises to the west because it is on the rim and ejecta of the NSM (see Chap. 6). A radial profile of the Orientale Basin is shown in Fig. 2.8.

2.3.5 Megabasins

This term is introduced (Byrne 2006) for an impact feature whose diameter is of the order of the South-Pole Aitken Basin (SPA, 1,500 km). The prefix mega- can mean either very large or one million. In terrestrial geology, megabasin is used for

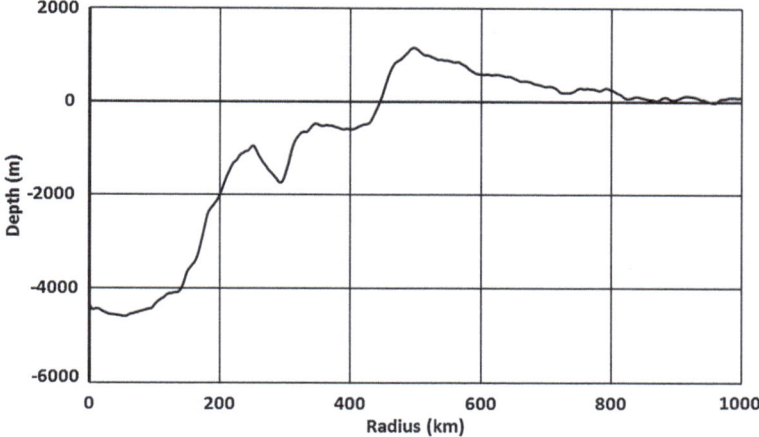

Fig. 2.8 Radial profile of the Orientale Basin. The main ring (Montes Cordillera) and two inner rings (inner and outer Montes Rook) are clear in the radial profile. A third ring inside of the Montes Rook and arcs of two outer rings are too subtle to appear in the radial profile. *Source* JAXA, Kaguya, 1/16° DEM, current author

Fig. 2.9 Elevation map centered on the South Pole-Aitken Basin (Lambert equal-area projection, 90° range, 180° central meridian). SPA (major apparent diameter 1,500 km) is a megabasin since there are other basins within it. Its rim extends from the South Pole to nearly the center of the far side bulge. *Source* JAXA, Kaguya, false color by the current author (scale in Fig. 1.2)

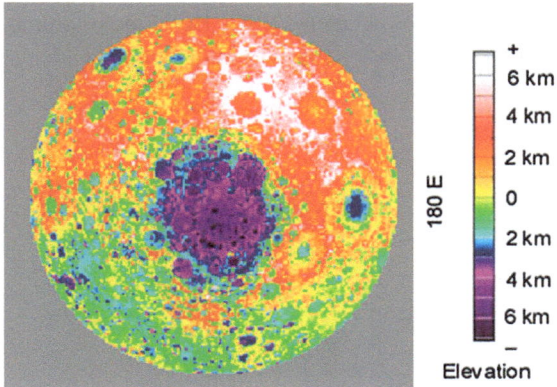

a watershed or ocean floor that contains smaller basins. In Mars geology (Minton et al. 2012) it is used for basins whose apparent diameter is over 1,000 km, in the spirit of the 300 km limit for basins. Here, we use that definition with tolerance for impact features that may be a little smaller but that have similar properties as other megabasins. An elevation map of the SPA megabasin is shown in Fig. 2.9. Figure 2.10 shows a radial profile of a model of the SPA megabasin as it would have appeared if the far side bulge were not there. Parameters of the SPA are described in Chap. 7.

Fig. 2.10 Radial profile of a model of the SPA megabasin. *Source* Chap. 7

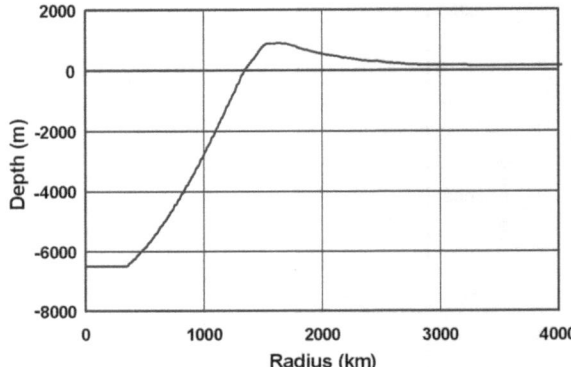

Such giant basins have been found by recent three-dimensional simulations (Ivanov 2007; Stewart 2011) to develop columns of melted and vaporized material far below the apparent crater. The melt column, which also contains pressurized vapor, rebounds above the apparent crater floor. As it falls back and cools, it forms a level floor in the apparent crater. Initially, the material of that floor may be chaotically mixed crust and mantle material but as it cools and solidifies, there is a re-crystallization and the surface material becomes predominantly crust. Flooding by mare basalt may follow much later. The term melt cylinder has been used for this phenomenon (Ivanov 2007). Here, melt column is used to reflect the slightly irregular shape, especially for oblique impacts.

Chapter 3
Radial Elevation Profiles of Large Craters and Basins Establish Self-Similarity

3.1 Introduction

Radial elevation profiles of impact features are found to be self-similar if scaled separately by diameter and depth. Following exploration of the Moon by the Apollo missions, the shape of the Moon was seen to be dominated by circular features created by hypersonic impact. Apollo altimeter data also revealed the far side bulge. An early proposal by Wood (1973) suggested that that bulge resulted from the ejecta from near side craters. This suggestion was pursued by Cadogan (1974) and Whitaker (1981), who proposed the Gargantuan and Procellarum Basins, respectively, as possible near side basins that may have produced the far side bulge. With the availability of Clementine data, both proposals were found wanting (Neumann et al. 1996) but the possibility remained that some other basin could explain the far side bulge. Previous attempts had used intuition and topographic hints for proposals.

   Here, another type of investigation is followed. Starting with the Maxwell Z-model (1977) and published extensions and comparing that body of theory with empirical observations of large lunar craters, further extensions were made that successfully produced an algorithmic model of lunar impact features over a wide range of sizes. Recently, 3-D simulations of impacts presented new information about the very largest impacts, the megabasins like the South Pole-Aiken Basin, that explained some of the empirical observations.

   There are two related methods of analyzing impacts. One is observing many impact events in the field (Earth and Moon) and in the laboratory. Then the observations can be classified and fit to algorithmic laws based on the Maxwell Z-model. The second method is to create computer simulations based on the state equations of expected materials. The state equations are also based on observations and the results of the computer programs are compared with the observations of impact features. Both methods have contributed to the understanding of impact craters. Extensions to the Maxwell Z-model are useful in describing the conversion of shock wave energy to kinetic energy, while simulation based on a state equation describes the effects of conversion of shock wave energy to phase changes in the target material.

C. J. Byrne, *The Moon's Near Side Megabasin and Far Side Bulge*,
SpringerBriefs in Astronomy, DOI: 10.1007/978-1-4614-6949-0_3,
© Charles J. Byrne 2013

3.1.1 Radial Elevation Profiles

This chapter describes observations of the large lunar craters and basins and establishes scaling laws that apply to the apparent craters of all nearly circular hypervelocity impact features. A new method of observation of large lunar features (Byrne 2006) is based on analysis of the newly-available digital elevation maps (DEMs) from spacecraft missions such as Clementine, Kaguya, and Lunar Reconnaissance Orbiter.

Radial elevation profiles are much more effective for circular features than linear profiles because they average out chaotic variations, making use of much more information. These profiles are used here to derive canonic structures of lunar events, reducing measurement errors and distortion by neighboring features, both before and after an impact event.

Earlier radial elevation profiles (Byrne 2007) were based on the Clementine 1° or 1/4° DEM but the ones reported here are based on the Kaguya 1° or 1/16° DEM. The increased coverage and resolution of the Kaguya data led to the discovery of several new features, leading to a complete model of the Moon covering features larger than 200 km in diameter.

3.1.2 Normalization by Both Diameter and Depth

A specific set of impact features is selected to establish a new scaling rule, normalization by both diameter and depth of an apparent crater that establishes the self-similarity of all apparent craters, large and small. This principle established a method to extrapolate to even larger craters, the megabasins. The Maxwell Z-model, that assumes that all shock wave energy is converted to kinetic energy, predicts that apparent craters will have a constant depth to diameter ratio. This is true for small craters (less than 10 km in apparent diameter, on the Moon) but is not true for larger craters that become progressively shallower with size, as more of the shock wave energy is converted to heating and phase changes in the target material. Allowing the model of an apparent crater to be scaled by both depth and diameter permits its use to be extended to craters of any size. Other extensions to the Maxwell Z-model (Maxwell 1977) that apply to these large features are discussed in Chap. 4.

3.1.3 Applications of the Model: Finding the Near Side Megabasin and Establishing a Complete Large Scale Model of the Moon

The new model was applied to the large craters, the basins, and the megabasins; and led to the successful search for the near side megabasin (NSM), the basin whose ejecta created the far side bulge and established the remarkable history of the near side. The extended Maxwell Z-model algorithms were used in the search for a new large crater by varying the parameters of the model (center coordinates, radius, depth, etc.) and searching for the combination that would minimize the standard deviation of the differences between the resulting model and the Clementine DEM. The result was the discovery of the NSM (Byrne 2007).

The addition of the NSM and refinement of the parameters of the South Pole-Aitken Basin led to the discovery of several new features (described in Chap. 5) by examining the residual DEM computed by subtracting the progressively improved model from the current topography.

The remainder of this chapter is a description of the doubly normalized model of impact features.

3.2 Radial Elevation Profiles Relative to Target Surface Shape

Radial elevation profiles were measured for a representative set of lunar impact features to see if similarities in their shapes could be found. Here, the measurements were based on the Kaguya 1/16° DEM. The selected features for the measurements are the Tycho and Grimaldi craters and the Humboldtianum, Humorum, Korolev, and Orientale basins. A detailed example of the process will be shown for the Orientale Basin. It was apparent that there were two issues to be addressed.

- Is the feature nearly circular?
- How is the shape of the apparent crater influenced by the target surface?

To address these issues, the radial elevation profile was computed for four quadrants, (NE, NW, SW, and SE).

The first step in the computation is to assign initial values of the latitude and longitude at the center of the feature (usually from literature estimates or examination of a map) and calculate the average elevation profile over each of the four quadrants. Then the center coordinates are adjusted to produce as symmetric quadrant profiles as possible, especially for the peak of the rim and the slope leading up to the rim (see Fig. 3.1). The depth at the center of the apparent crater cannot usually be directly observed because it is obscured by fill.

The shapes of these features are usually distorted by neighboring features, both before and after impact. The slopes of these large features are small for the most part, of the order of a few kilometers vertically over more than 100 km horizontally. For such low slopes, the shock wave, excavation cavity, and apparent crater are assumed to follow the target surface (the principle of superposition). Therefore, estimation and subtraction of the target surface from the radial profile will reveal the approximate canonic shape of the impact feature and permit measurement of parameters such as the apparent diameter and apparent depth that are relative to the target surface. The object is to estimate what the feature would have looked like if it had impacted a flat target, so as to permit comparison of the parameters over different impact events. The components of the pre-impact surface that are estimated are elevation at the center of the apparent crater, slope and saddle, and linear and curvature

Fig. 3.1 The elevation of the Orientale Basin is averaged over each of four quadrants of azimuth as a function of radius. The quadrants are: *solid, NE, dash dot dash NW, dash SW,* and *long dash, SE*. The center coordinates are adjusted to produce the strongest average signal for the rim and slope up to the rim. *Source* Kaguya 1/16° DEM

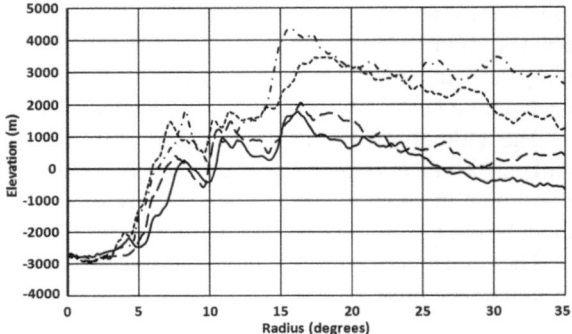

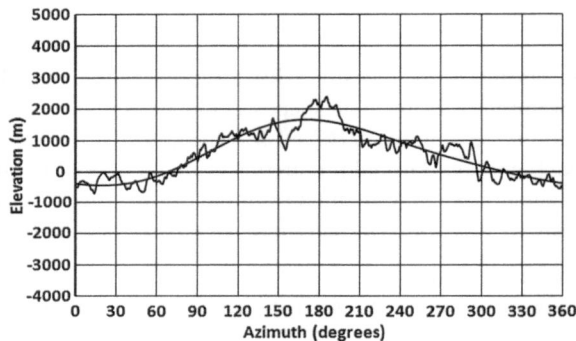

Fig. 3.2 The DEM of the Orientale Basin is averaged over radius as a function of azimuth (*black*). The *curve* is the sum of the fundamental and second harmonic Fourier terms, an estimate of the slope and saddle components of the target shape. The average of the fundamental and harmonic elevations over each quadrant is taken. The quadrant averages, multiplied by the radius, are subtracted from the radial profiles for each quadrant

radial components (note that the overall curvature of the Moon is removed in the construction of the DEM). The shape of the target surface z_T to be estimated is:

$$z_T = a + br + c \cdot r^2 + r \cdot [d \cdot Sin(\theta) + e \cdot \cdot Cos(\theta)$$
$$+ f \cdot Sin(\theta/2) + g \cdot Cos(\theta/2)] \tag{3.1}$$

where z_T is the target's elevation and r, θ are the polar coordinates from the center.

To estimate slope and saddle, the average elevation (over a radius of about twice the radius of the cavity) is calculated as a function of azimuth (see Fig. 3.2). Coordinates d, e, f, and g as functions of azimuth are automatically calculated.

The program averages the first and second harmonics over each quadrant and subtracts them from the corresponding elevation profiles, allowing the slope and saddle shapes to be removed (see Fig. 3.3).

Fig. 3.3 These are the four quadrant profiles of the Orientale Basin after slope and saddle have been removed. Note that the profiles for the four quadrants are in closer agreement than in the raw profiles of Fig. 3.1

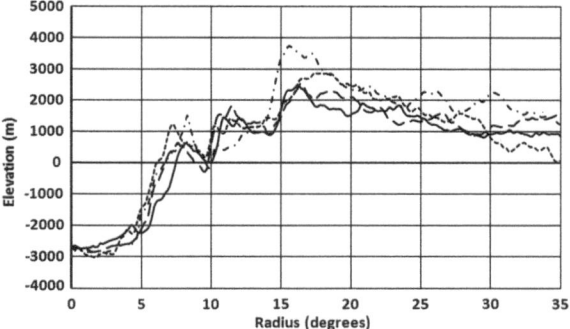

Fig. 3.4 The radial profile of the Orientale Basin after adjusting for slope and saddle (the average of the four profiles of Fig. 3.4). The parameters of the *smooth curve* are the remaining estimated parameters of the pre-impact target surface

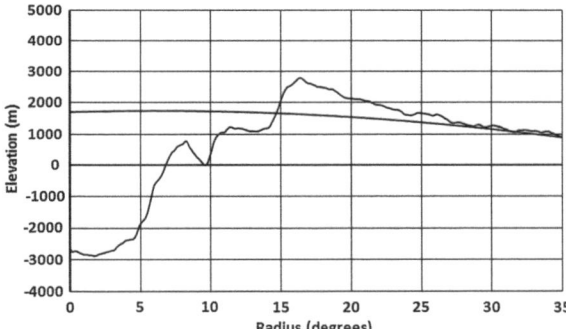

After removing the slope and saddle, the four quadrant profiles are averaged to produce a profile that is independent of the d, e, f, and g terms of the target surface. If necessary, specific quadrants are suppressed because of strong interference of a neighboring feature (this is not needed for the Orientale Basin).

The remaining parameters, elevation, linear, and curvature (a, b, c) are estimated manually (see Fig. 3.4). After removal, the result is the approximate shape of an apparent crater as it would have been if the impact had occurred on a flat surface (see Fig. 3.5).

Judgment is needed only in the estimation of the latitude and longitude of the center of the feature plus the elevation and the radial linear and curvature components of the target surface. The other corrections are made automatically.

3.2.1 Selected Features, Normalized to Show Self-Similarity

The next step is to compare the radial profiles of the selected features in order to establish a prototypical shape for a model. For these larger features, it is necessary

Fig. 3.5 This is what
the apparent crater of the
Orientale Basin would have
looked like if the target
surface had been flat

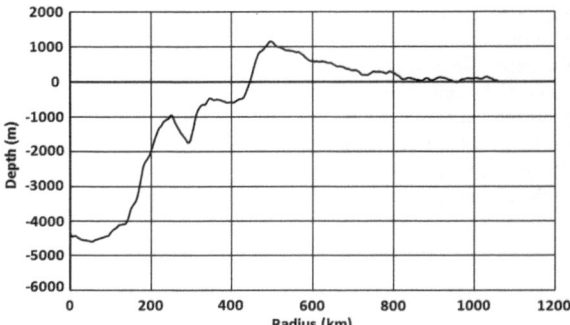

Fig. 3.6 This *graph* shows
the doubly normalized radial
profiles of two complex
craters and four basins
covering a diameter range
of 20:1. The *elevation 0*
represents the equivalent flat
target surface. The *radius 1.0*
(normalized apparent radius)
is at the intersection of the
cavity with the target surface

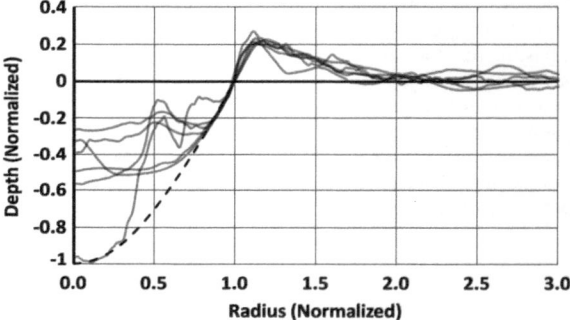

to normalize both the radius by the apparent diameter and the depth by apparent
depth, as estimated separately.

The result of double normalizing the radial elevation profiles of the selected
features is shown in Fig. 3.6. All of the cavity profiles are observed to be bounded
by the dashed curve that envelopes their diverse features of fill and internal rings.
That curve is modeled by:

$$z/d_a = -Cos\,[(r/R_a) \cdot (\pi/2)]$$

(3.2)

where z is the elevation, d_a is the apparent depth, r is the radius from the center,
and R_a is the apparent radius.

The shape of a simple crater (Linné, 2.22 km) has been modeled by an exponential curve (Garvin et al. 2011). An exponential curve could also be used here,
but would be very similar to the cosine curve. Until a physical explanation is available to distinguish between the two approaches, I favor the cosine curve because it
has one less parameter than the exponential curve.

The ejecta fields, rims, and slopes up to the rims in Fig. 3.6 are self-similar.
That is, their shapes are independent of size in the observed range. This implies
that they can be described by algorithms that can be applied once the diameter

and depth of the apparent crater is estimated. The internal cavities are somewhat diverse although they all are bounded by an envelope similar to the bowl shape of a simple crater. The peak-rings show approximate similarity. The model of a peak-ring is a raised cosine superposed on the envelope curve.

The floors of the craters do not exhibit similarity. They may have been partially formed shortly after the impacts but also could be caused by a combination of mare flooding, ejecta from nearby impact features or pyroclastic deposits.

The similarity of the ejecta fields on this doubly-scaled graph shows that the volume of the ejecta scales with the product of the apparent depth and the square of the apparent diameter. Quantitatively, the average volume of the ejecta fields are approximately the same as the volume of the dashed curve of Fig. 3.6. This suggests that the dashed curve represents the apparent crater and the individual features within that curve represent rebound and subsequent modification processes.

These craters and basins cover a wide range of apparent diameters, from 74 (Tycho) to 950 km (Orientale Basin). The inferred scaling law applies to smaller craters, even below 10 km where the depth to diameter ration is constant, since scaling separately by depth and diameter in that region is equivalent to scaling by either one. Further, it is established in Chap. 5 that the scaling applies to the megabasins as well, even up to the size of the Near Side Megabasin, whose cavity covers more than half of the Moon. An algorithmic model for the self-similar parameters of the selected features is developed in Chap. 4.

Note that the self-similarity scaling rules apply only to the topography, formed by kinetic energy impacted to target material by the shock wave, and does not apply to matters involving heating and phase changes, including the formation of the melt column beneath megabasins.

Chapter 4
A Model of Large Impact Craters and Basins: Their Cavities, Rims, and Ejecta Deposits

4.1 Introduction

The Maxwell-Z model of impact features is expanded from later additions in the literatures and additionally expanded here to model lunar impact features greater than 10 km in diameter; the model is used to fit 74 impact features. Maxwell developed the first model of a crater caused by a hypersonic shock wave launched by a surface explosion (Maxwell 1977). Croft extended the Maxwell Z-model to cover subsurface explosions, since hypervelocity impactors penetrate the surface before transferring most of their energy to a shock wave (Croft 1980). Housen, Schmidt and Holsapple further extended the Maxwell Z-model to introduce dimensional modeling and produce equations for a wide variety of variables, including the launch velocity of ejecta (Housen et al. 1983). After these extensions, the Maxwell Z-model was very successful in describing the shape of hypervelocity impact events in the laboratory, on Earth, and on the Moon, for features up to 10 km in diameter. However, there was a puzzle. The Maxwell Z-model predicted that the depth-to-diameter ratio of a crater in a particular target material was constant with the size of the crater. Yet observation found that craters beyond a certain size (10 km in diameter for the Moon) became progressively shallower with size, not only on the Moon but on Earth, Mars, and Mercury as well (Hiesinger 2006; Sharpton 1994) (see Fig. 4.1).

One cause for this effect is that large features impacted earlier than 3.8 Ga, when the Moon may have been still somewhat plastic and allowed isostatic compensation. Further, radioactive heating of the crust and upper mantle as late as 3.5 Ga may also have promoted compensation. However, isostatic compensation, which certainly occurred, would not have fully explained the progressive shallowness of the largest features. The physical cause of the progression of shallowness is not fully understood; it may be related to the increasing diversion of shock energy from imparting kinetic energy to heating and phase changes. However, building on the principle of normalizing by depth and diameter separately as described in Chap. 3, it is now possible to scale an impact model over the full range of diameters. This principle may also be useful for impact features of other rocky bodies.

C. J. Byrne, *The Moon's Near Side Megabasin and Far Side Bulge*,
SpringerBriefs in Astronomy, DOI: 10.1007/978-1-4614-6949-0_4,
© Charles J. Byrne 2013

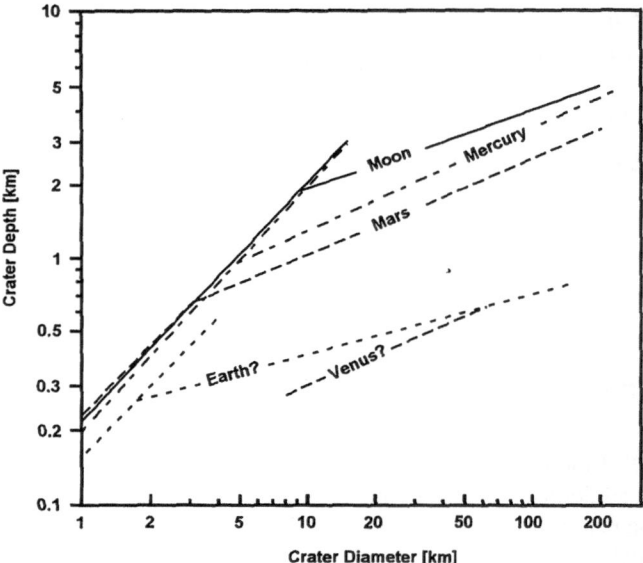

Fig. 4.1 The depth to diameter ratio of craters smaller than a certain size is a constant, as predicted by the Maxwell Z-model. Above a break point 10 km for the Moon, the ratio is constant, but decreases as size increases (Hiesinger and Head 2006; Sharpton 1994). *Source* (Hiesinger and Head 2006)

Extrapolating the extended Maxwell-Z model was used in the search for a new large crater by varying the parameters of the model (center coordinates, radius and depth) and searching for the combination that would minimize the standard deviation of the differences between the resulting model and the Clementine DEM. The result was the discovery of the Near Side Megabasin (NSM) (Byrne 2007).

As the Kaguya 1° and 1/16° DEMs became available, the extended model was applied to them, with increased accuracy, especially for the polar zones. As a result, the parameters of the NSM and the South Pole-Aitken Basin were refined and it was possible to find the parameters of all features greater than 200 km in diameter. The remainder of this chapter is a description of the model of impact features that was used in this exercise.

4.2 The Maxwell Z-Model

There are three central principles to the Maxwell Z-model, based on Maxwell's observations of surface explosions (Maxwell 1977).

- After the passage of a shock wave, a particle in the excavation zone has a radial velocity from the center of the explosion given by:

$$dR/dt = \alpha \cdot R^{-Z}. \tag{4.1}$$

- Although the shock wave transferred energy to each particle by compression as it passed, the ejection process is assumed to be incompressible.
- After a particle is ejected from the target surface it follows a ballistic trajectory and is deposited in the ejecta field.

These three assumptions are sufficient to determine the nature of the ejection process if α (related to the energy of the explosion) and Z (a property of the target material) are assumed constant. The result agrees approximately with observations over a wide range of scales from laboratory experiments to lunar craters less than 10 km in diameter.

The flow proceeds along streamlines spiraling out from the center of the explosion, first downward, then outward, and then upward, emerging at an angle of about 45°, with a speed that decreases as the radius of the point of ejection increases, until ejection terminates near the edge of the transient crater, where the shock wave becomes an acoustic wave. Figure 4.2a is a useful in visualizing the flow field.

The material that is ejected from the feature is bounded above by the target surface and below by a surface whose cross section is that streamline that extends from the effective center of the shock wave to the intersection of the target surface with the rim. This circular intersection is also at or near the top edge of the transient crater, where the shock wave becomes a sonic wave, ending the excavation process.

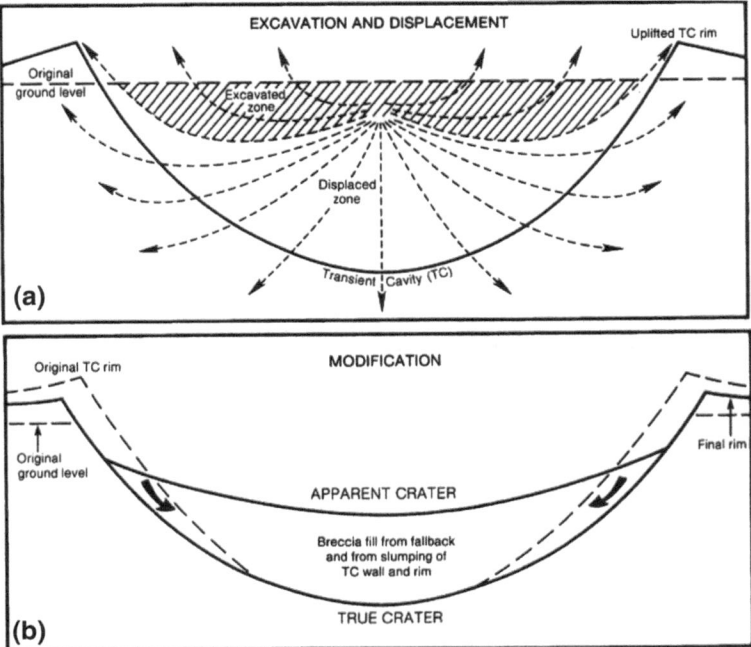

Fig. 4.2 **a** Transient cavity and excavation cavity of a simple crater. **b** Apparent crater, formed by the collapse of fractured target material. *Source* Lunar Sourcebook (Heiken et al. 1991)

The Maxwell Z-model assumed that the shock wave, as it passed through the target material, imparted energy to the target material. Subsequently, that energy was released in the form of kinetic energy for work against strength, gravity and inertia. For the large craters and basins considered here, the energy devoted to overcoming strength can be neglected. The result is to eject material from the transient cavity. The lower boundary of the excavation zone is defined by the streamline that ends at the edge of the crater rim, called the hinge streamline. The term "hinge" refers to the ejecta field having its upper layer derived from lower layers of the original target material, effectively rotating around the point terminating the hinge streamline.

There are very important limitations to the Maxwell Z-model, in addition to the assumption that the shock wave is launched from the surface. There is evidence that the value of Z is not constant but can vary between two directly below the center of the shock wave to four at the surface (Croft 1980). In addition, there is no place in the model for heating, melting, and vaporization in the target material. All the energy imparted to the material by the shock wave is assumed to be converted to kinetic energy, in an initial velocity.

We need to establish the normalized algorithm that controls the velocity profile early, because the subsequent evolution of the rim and ejecta blanket depends upon it. Once the normalized ejection launch velocity profile is established, new algorithms need to be established for the process of transporting the ejecta to the range of deposit. Finally, we need to examine the algorithm that converts the rate of volume ejecta to the depth of the rim and ejecta blanket.

4.3 The Maxwell Z-Model Applied to Impacts

An impactor penetrates the target surface as it releases its kinetic energy to the shock wave launched in the target. In the process, one shock wave is generated in the target and another is generated in the impactor. The effective center of origin of the shock wave and the subsequent effective center of the ejecta flow is beneath the surface at a distance d that is dependent on the physical size of the impactor. Observations indicate that the depth of this effective center is typically about equal to the diameter of the impactor (Croft 1980). It follows that the Maxwell Z-model flow takes time and space to be established. The model does not precisely apply to this phase of the entry of the impactor.

Figure 4.2a shows the formation of a transient cavity, the subsequent excavation and displaced zones, and the launch of ejecta. Figure 4.2b shows the collapse of material immediately after the ejection process completes, forming the apparent crater.

These assumptions allowed Croft (Croft 1980) to present a set of equations for the shape of the streamlines, maximum depth of excavation, the angle of ejection at the target surface, and the ejecta volume. The notation for the equations is shown in Fig. 4.3. The parameters of the impact crater were reduced to its apparent diameter and the apparent depth of the crater, data that can be easily measured for young simple craters. These parameters must be estimated for older, larger

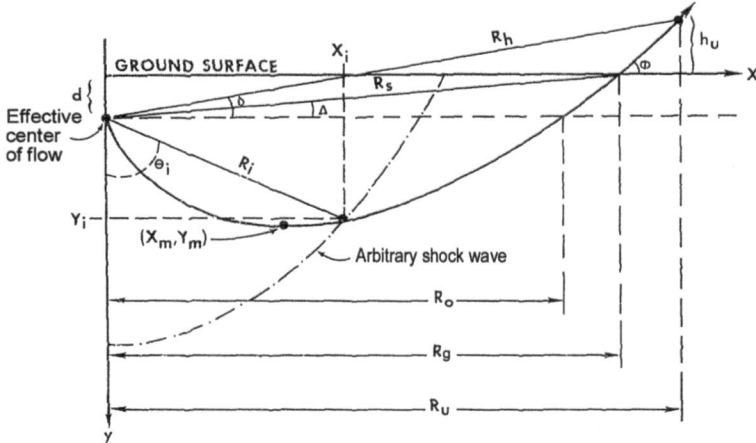

Fig. 4.3 A shock wave (*dashed line*) originating at a *depth d* and a resulting streamline (solid curve) are shown. The hinge streamline intersects the ground surface near the rim of the apparent crater at $R_g = R_a$ (the apparent radius). It is called the hinge streamline because the top of the rim and the ejecta field invert the depth profile of the target material, as if a flap of material has been turned around the end of the hinge streamline. *Source* (Croft 1980)

craters which have been degraded. The apparent depth is often particularly diffi-cult to estimate because the bottom of a crater is often obscured by a central peak, or by mare, or by ejecta from nearby features.

The notation in this figure will be used (with additions) in the remainder of this chapter.

The following two equations are from the appendix of (Croft 1980). The equa-tion for each streamline is:

$$R_i = R_0 \cdot (1 - \cos(\theta_i))^{1/(Z-2)}.$$ (4.2)

The bottom of a streamline is:

$$Y_m = \frac{X_i \cdot (Z-2) \cdot (Z-1)^{(1-Z)}}{\cos(\Delta) \cdot (1 - \sin(\Delta))^{1/(Z-2)}} + d.$$ (4.3)

Assuming Z is approximately constant and Δ is small and that the average excavation depth is proportional to the maximum excavation depth, the average excavation depth of a streamline is:

$$Y_a = K \cdot X_i + d$$ (4.4)

where K is represents the function of Z and Δ within Eq. (4.3).

If we take the depth of excavation of the transient crater to be approximately the depth of the apparent crater, then the depth to diameter ratio will be constant, as is observed for lunar craters under 10 km in diameter. The relative shallowness of large craters and basins is evidence that something must change in the Maxwell Z-model for large craters and basins.

If the value of Z is reduced from its typical value of about 3 toward 2.0, then, according to Eq. (4.2), the hinge streamline becomes progressively shallower but the angle of ejection is reduced accordingly. Perhaps the constant Z should be replaced by a variable, a function of depth, local gravity, and density that combine to establish an ambient pressure.

Unfortunately, we do not have such a function so another type of analysis will be used, one appropriate to a very large class of self-similar phenomena: dimensional analysis. Croft's insight that the effective center of the shock wave is at a depth d will be used in the context of dimensional analysis as part of an energy balance equation for determining the normalized ejection velocity profile. The qualitative aspects of streamlines will also be retained.

4.4 Dimensional Analysis

Dimensional analysis is a field that develops the forms of equations and isolates dimensioned variables from dimensionless parameters of the equations. The parameters are determined by observations, which also serve to establish scaling laws and to establish the scope where the parameters of the equations are at least approximately constant. The critical assumption is that the phenomenon being examined is capable of scaling, at least over a range of sizes. The field was developed for hypervelocity impact features by Housen, Schmidt, and Holsapple (Housen et al. 1983; Holsapple et al. 1982). The resulting model extended the Maxwell Z-model to cover the velocity of ejection and the shape of the ejecta blanket.

A note of caution is called for here. The empirical data presented in Chap. 3 concerning lunar craters more than 10 km in diameter show that horizontal measurements scale by apparent diameter and vertical measurements scale by apparent depth. So we need to be cautious about simply scaling by length if we do not recognize the difference in scaling by direction.

In dimensional analysis, the focus is on ejection conditions as a function of radius, not on the streamlines. The velocity of ejection, as a function of internal radius determines the range to the radius of deposit. The rate of ejecta volume per increment of internal radius (also the rate of material deposited) and the increment of range determines the depth of the deposit.

Once the form of the equations for velocity, range, and volume rate are determined, the parameters are adjusted to meet the observed shape of the rim and ejecta field shown in Fig. 3.6. The first step is to determine a function for velocity.

There are two major regimes in cratering mechanics, the strength regime and the gravity regime. For large craters and basins, the energy needed to overcome the strength of the material is negligible so the equations for the gravity regime are used.

A scaling relation for ejection velocity as a function of launch position was developed (Housen et al. 1983):

$$v/\sqrt{gR_a} \propto (X_i/R_a)^{(\alpha-3)/2\alpha}. \tag{4.5}$$

where v is the magnitude of the launch velocity, g is the acceleration of gravity, R_a is the apparent radius, α is a property of the target material, and other symbols are as defined in Fig. 4.3.

Equation (4.5) does not consider the energy converted to potential energy as the incremental volume of each streamline is raised to the surface, a discrepancy corrected in the next section.

4.5 A New Term Added to the Launch Velocity

In a major advance, J. E. Richardson provided new insight to the ejection velocity analysis, by including a term in an energy balance equation for the energy needed to raise the ejecta to the surface (Richardson 2007). Restating (4.5) as:

$$v^2 = C_{vg}^2 \cdot g \cdot R_a \cdot (X_i/R_a)^{2/\mu} \tag{4.6}$$

where C_{vg} and μ are target material constants. The parameter μ replaces and simplifies the exponent of (4.5):

$$\mu = 2 \cdot \alpha/(\alpha - 3) \tag{4.7}$$

The variables of (4.6) are defined in Fig. 4.3.

The contribution of Richardson is to apply an energy balance equation. The kinetic energy per incremental volume of ejecta at the surface is equal to the total kinetic energy of the stream line less the increase in potential energy as the streamline is raised to the surface;

$$^1/_2 \cdot \rho \cdot v_e^2 = ^1/_2 \cdot \rho \cdot v^2 - \rho \cdot g \cdot Y_a \tag{4.8}$$

where ρ is the ejecta density.

Divide by $1/2 \cdot \rho$, substitute for v^2 from Eq. (4.6), and substitute for Y_a from Eq. (4.4):

$$v_e^2 = C_{vg}^2 \cdot g \cdot R_a \cdot (X_i/R_a)^{2/\mu} - 2 \cdot g \cdot (K \cdot Xi + d) \tag{4.9}$$

Note that the contributions of Croft, the location of the center of the shock wave below the surface, and Richardson, the inclusion of potential energy, have been combined in Eq. (4.9). The concept of streamlines has been retained from Maxwell, but the shape is not fixed, nor does all of the shock energy transform to kinetic energy. Dimensional analysis is a higher order of modeling that relies on empirical evidence of self-similarity.

4.6 Normalized Velocity Equation

The shape of the ejection field for the large craters, larger than 10 km but smaller than the megabasins, can be scaled from three normalized equations. The three (all functions of the normalized apparent diameter, normalized apparent depth, and normalized internal ejection radius) are for the ejection velocity, range, and incremental volume.

The square of the ejection velocity is normalized by dividing by $g \cdot R_a$:

$$v_{eN}^2 = v_e^2/(g \cdot R_a) \tag{4.10}$$

Then, from Eq. (4.9):

$$v_{eN}^2 = C_{vg}^2 \cdot (X_{iN})^{2/\mu} - 2 \cdot K \cdot X_{iN} - 2 \cdot d_N \tag{4.11}$$

where $X_{iN} = X_{i/Ra}$ and $d_N = d/R_a$.

This is the first of the three normalized algorithms to establish the rim and ejecta field shape. The next algorithm to be sought is the one for range of the ejecta.

4.7 Normalized Range Equation

The ejecta field is deposited at a ballistic distance X_d from the launch point at X_i. For the large craters and basins, the target surface can be assumed to be approximately flat and the trajectory a parabaloid. The error becomes too large for the megabasins, the radius of the Moon becomes an important parameter, and the resultant range equation (see Sect. 4.12) cannot be normalized. However, the parameters that can be derived by matching the empirical self-similar rim and ejecta field can be applied to the megabasins as an extrapolation.

For a parabolic trajectory, the radius of deposit of ejecta is:

$$X_d = X_i + 2 \cdot (v_e^2/g) \cdot \sin(\Phi) \cdot \cos(\Phi) \tag{4.12}$$

where Φ is the launch angle (in radians).

Normalizing the launch radius and range on R_a and v_e^2 on $g \cdot R_a$ as before:

$$X_{dN} = X_{iN} + 2 \cdot v_{eN}^2 \cdot \sin(\Phi) \cdot \cos(\Phi) \cdot \tag{4.13}$$

This is the second of the three normalized algorithms we seek. The last algorithm to be found is the one that controls the normalized depth of the deposited ejecta.

4.8 Normalized Depth Equation

The incremental volume of material ΔV_i ejected from the increment of radius ΔX_i is calculated from the scaling function for volume V_e, ejected within the launch point X_i (Housen 1983).

From (Housen et al. 1983), modified to allow for the double normalization established in Chap. 3:

$$V_e/(d_a \cdot R_a^2) \propto (X_i)^3/(d_a/R_a^2) \qquad (4.14)$$

where d_a is the depth of the apparent crater.

Normalizing (4.14) on $d_a \cdot R_a^2$, taking the derivative to find the finite difference relationship, and assigning a constant of proportionality:

$$\Delta V_{eN} = C_V \cdot (X_{iN})^2 \cdot \Delta X_{iN} \qquad (4.15)$$

where X_{iN} and ΔX_{iN} are normalized on R_a.

The depth of the rim and ejecta blanket is the incremental ejected volume divided by the width and circumference of the incremental deposit ring:

$$D_N = D/d_a = \Delta V_{eN}/(\Delta X_{dN} \cdot 2 \cdot \pi \cdot X_{dN}) \qquad (4.16)$$

where X_{dN} and ΔX_{dN} are normalized on R_a.

Substituting for ΔV_{eN} from Eq. (4.15):

$$D_N = C_V/(2 \cdot \pi) \cdot (X_{iN})^2 \cdot \Delta X_{iN}/(\Delta X_{dN} \cdot X_{dN}) \qquad (4.17)$$

Regrouping similar terms:

$$D_N = [C_V/(2 \cdot \pi \cdot X_{iN})] \cdot (X_{iN}/X_{dN}) \cdot (\Delta X_{iN}/\Delta X_{dN}). \qquad (4.18)$$

Equation (4.18) is the last of the three equations needed to produce a finite difference computation of the normalized rim and ejecta field.

4.9 Velocity Profile, Target Parameters, Rim, and Ejecta Field Models

The three normalized equations were calculated, with a normalized radius increment of 0.01. With appropriate values of target material constants found by trial and error, a very good fit resulted between the calculated elevation profile of the ejection field beyond a value of normalized deposit radius of 1.3, corresponding to a normalized launch radius of 0.93. However, the calculated depth of the rim (within those values) was much too large. Clearly, the formation of the rim does not match the normalized equations derived above. The likely problem is that the rim is partly formed by the deposit of ejecta and partly by uplift of the crater lip, a common observation of impact craters that unfortunately has no clear model as yet. Modeling the uplift process would be a good project for the future.

After modifying the velocity profile by trial and error, it turns out that the observed rim can be approximated by assigning values to the detailed velocity profile in the range of 0.94 and 1.0 of the apparent radius. For the range of 0.99–1.00, it was necessary to reduce the increment of radius from 0.01 to 0.002. The new values of the detailed velocity profile are shown in Table 4.1 and the modified velocity profile is shown in Fig. 4.4.

The values of the parameters of the target surface that provide a best fit to the rim and ejecta field of the selected features are shown in Table 4.2 and the resulting rim and ejecta field is shown in Fig. 4.5.

Table 4.1 Normalized values for the modified velocity profile

Radius	0.94	0.95	0.96	0.97	0.98	0.99
Velocity	0.709	0.664	0.614	0.556	0.485	0.392
Radius	0.992	0.994	0.996	0.998	1.0	
Velocity	0.368	0.335	0.292	0.230	0.000	

Table 4.2 Constants for the ejecta equations

Gravity	Ejection Velocity					Depth
$g \cdot m/s^2$	C_{vg}^2	μ	K	d_N	Φ°	C_v
1.62	0.73	−0.2	0.44	0.038	45	0.82

Notes d_N is the ratio of d, the effective depth of the center of flow, to R_a, the apparent radius. Depth changes the depth of the rim and ejecta blanket without changing the shape. Gravity is the value of the Moon's surface gravity, and can be changed for any other body

Fig. 4.4 The normalized velocity profile that provides a good match to the ejecta field profiles of Fig. 3.6. The corresponding profile of the rim and ejecta field profile derived from this velocity profile is shown in detail in Fig. 4.5 and against the set of selected large craters and basins in Fig. 4.6

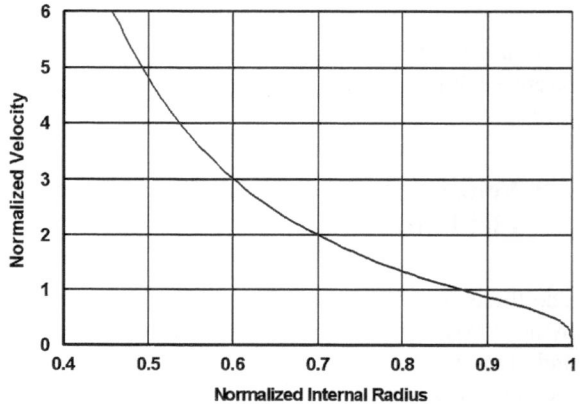

Fig. 4.5 The doubly normalized model of the rim and ejecta field produced by the modified normalized velocity profile

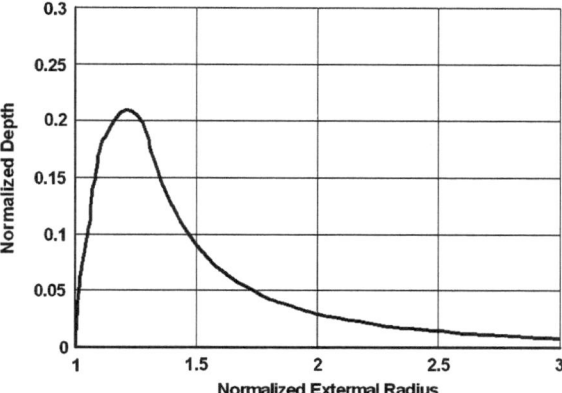

The full set of models is shown in Fig. 4.6 and compared to the radial profiles pf tje set of selected features. The double-normalized volume, directly calculated from the external profile of the model, is found to be 1.60. The volume of the envelope of the internal cavity (the negative cosine model) is 1.45. In other words, the ejecta volume is about 10 % greater than the internal volume of the space it came from, neglecting peak-rings and fill. This suggests three potential points:

- Rings and fill are due to processes that follow the ejection process.
- Ejecta and rims are more porous than the target material by about 10 %. An increase of porosity is to be expected because of fracturing and the packing density of boulders and blocks.
- The external volume beyond a radius of 3.0 is difficult to measure and may decrease more rapidly than predicted in the domain of rays and secondary craters.

4.10 Model of a Peak-Ring

For those complex craters that have a clear peak-ring, the observed amplitude and location tend to be quite consistent. A model that fits them well is a raised-cosine superimposed on the negative cosine model of an apparent crater. The height of the peak relative to the negative cosine model is D_{p-r}, centered at R_{p-r} and the width is $0.5 \cdot R_a$ centered at R_{p-r}. The equation for the model is:

$$Y_{p-r} = D_{p-r} \cdot 0.5 \cdot [1 + \cos([(X_i - R_{p-r})/0.25] \cdot \pi)] \qquad (4.19)$$

for the range of X_i from $R_{p-r} - 0.25 \cdot R_a$ to $R_{p-r} + 0.25 \cdot R_a$.

The normalized equation is:

$$Y_{Np-r} = D_{Np-r} \cdot 0.5 \cdot [1 + \cos[(X_{Ni} - R_{Np-r})/0.25] \cdot \pi)] \qquad (4.20)$$

for the range of X_{Ni} from $R_{Np-r} - 0.25 \cdot R_{Na}$ to $R_{Np-r} + 0.25 \cdot R_{Na}$ where the depth variables are normalized on D_a and the radial variables are normalized on R_a.

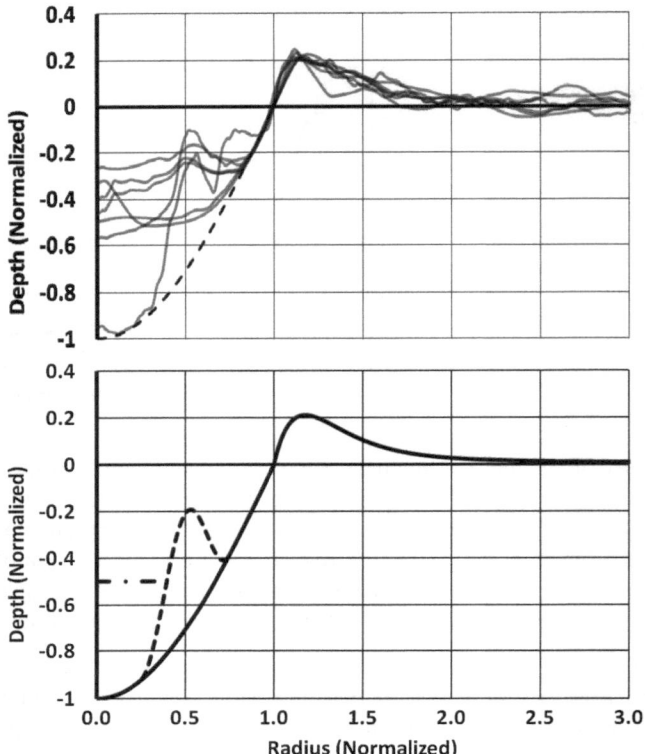

Fig. 4.6 The *upper graph* shows the double-normalized radial profiles of the 7 large craters and basins discussed in Fig. 3.6. Since features of these impact features are self-similar, they can be scaled from the normalized model shown in the *lower graph*. The *solid curve* models the apparent crater, rim, and ejecta field. The *dashed* line is a model of a peak-ring and the *dash-dot* line represents fill by maria or ejecta from other features

The values of D_{Np-r} and R_{Np-r} are usually close to 0.5 but they are measured for each impact feature that has a clear peak-ring. The model of a peak-ring is shown in Fig. 4.6.

4.11 Model of Fill

If the fill is by a mare, its surface will be approximately level relative to the geoid, like a lake or ocean on Earth. If the fill is by ejecta, it may be irregular, especially if the source is near. In either case, the fill will be approximated by a level surface as shown in Fig. 4.6. The depth of fill is measured from the center of the cavity to the level of the fill. In addition, the absolute elevation of the surface is measured.

Impact features are modeled in estimated historical order. As the feature is modeled, the horizontal extent of the fill is determined by comparison with the feature and stored. The fill is added after all models are superposed without fill. This process reflects the observation that both the mare fill and ejecta fill are usually added to a feature long after its impact. Accordingly, later features are superposed on an older feature's crater, rim or ejecta field before the fill is added. Most often (but not always), this is realistic.

4.12   Extension of the Range Equation for Megabasins

The agreement of the ejecta fields in Figs. 3.6 and 4.6 indicates that the ejecta approximately followed simple ballistic trajectories over the distances measured, up to three times the apparent radius of the features. To investigate larger features (the megabasins) this approximation is not valid; the range of each incremental cone of ejecta must be calculated for elliptical trajectories, starting at the launch radius of each streamline. The normalized velocity algorithm still holds in the megabasin domain, but the algorithm for the ejecta trajectory must take account of the radius of the Moon and cannot be normalized. Therefore the rims and ejecta fields of megabasins are not self-similar.

The equation for the range of ejecta deposit is given in (Thompson 1986). In our notation, it is:

$$X_d = X_i + 2 \cdot R_M \cdot \tan^{-1} \left(\frac{(v_e^2/R_M) \cdot \sin(\Phi) \cdot \cos(\Phi))}{(g - (v_e^2/R_M)) \cdot \cos^2(\Phi)} \right) \qquad (4.21)$$

where R_M is the radius of the Moon.

This equation could also be used for the large craters and basins.

4.13   Models of the Large Craters and Basins

A list of the impact features that have been modeled by the techniques of this chapter are shown in Table 4.3. All of the basins listed in The Geologic History of the Moon (Wilhelms 1987) were examined, but several of them could not be confirmed as impact features. That does not mean that they are proven not to be impact features, for they may have been so degraded by further impact or buried by deep mare flows that their topographic expression is subdued. A few impact features have been added (named in italics), identified typically by examination of newer datasets.

All impact features with apparent diameters of 200 km or more and confirmed to be impact features by this method are included in Table 4.3. As we shall see in the next chapter, there are no clear indications of missing impact features in

Table 4.3 Modeled impact features

Name	Period	Lat.	Long.	Diam.	Depth	Fill
		degree	degree	km	km	km
Imbrium	LI	34.8	−15.6	1,072	3.600	3.440
Lavoisier-Mairan	UI	42.1	−65.6	1,000	1.800	1.490
Orientale	LI	−19.6	−95.0	890	4.680	0.180
Nectaris	N1	−14.5	34.5	840	3.200	0.740
Crisium	N2	17.0	59.0	740	3.400	0.920
Nubium	pN3	−18.0	−15.0	712	3.200	2.700
Serenitatis	N2	27.4	19.5	696	3.600	2.800
Marginis	pN2	10.0	85.8	676	2.400	2.000
Fecunditatis	pN3	−1.8	52.0	660	2.800	2.100
Humboldtianum	N2	57.7	82.5	608	5.000	2.300
Flamsteed-Billy	N9	−5.0	−50.4	600	2.500	2.680
Mendel—Rydberg	N1	−49.6	−94.5	592	6.000	2.090
Moscoviense	N1	26.6	149.8	588	6.200	0.400
Cardanus—Herodotus	LI	21.8	−61.8	580	2.600	2.600
Freundlich-Sharanov	pN8	18.4	175.6	564	5.600	0.900
Hertzsprung	N2	1.8	−129.5	524	6.000	3.000
Apollo	pN9	−36.0	−151.4	480	6.800	4.180
Schiller-Zuchius	N	−55.5	−45.0	428	3.400	2.060
Kohlschutter-Leonov	pN4	13.6	155.0	420	3.400	2.400
Smythii	pN5	−1.6	87.4	420	6.000	4.100
Humorum	N2	−24.0	−39.0	420	3.200	2.240
Korolev	N1	−4.6	−157.4	398	9.000	5.100
Dirichlet—Jackson	pN	13.3	−158.0	384	6.800	5.100
Lorentz	pN6	33.6	−97.0	336	5.200	4.140
Mendeleev	N2	5.3	141.3	304	9.500	6.260
Coulomb—Sarton	pN5	52.4	−121.3	300	3.400	1.200
Schrödinger	LI	−74.8	133.2	298	8.000	6.100
Planck	pN7	−57.5	136.3	296	7.400	5.736
Poincarè	pN4	−57.0	163.0	292	3.600	2.750
Ingenii	pN4	−32.8	164.0	292	6.000	5.150
Birkhoff	pN7	59.0	−146.8	292	6.840	4.440
Fitzgerald—Jackson	pN	25.0	−169.0	276	3.600	2.100
Harkhebi	pN	40.2	99.0	250	6.400	5.000
Jules Verne	pN3	−35.3	147.0	244	4.200	2.120
Sikorsky-Rittenhouse	N2	−68.5	110.0	244	6.000	5.270
Sinus Iridum	UI	44.5	−31.2	236	6.200	5.580
Gagarin	pN4	−19.8	149.8	232	8.800	6.900
Milne	pN	−31.2	113.0	228	4.600	3.500
Von Karman M	pN	−46.6	175.7	224	4.200	3.700
Campbell	pN	44.5	154.3	216	7.000	4.000
Leibnitz	pN	−38.2	179.0	214	6.200	4.400
Jeans—Priestly	pN3	−58.6	101.8	212	2.600	1.300
D'Alembert	N	51.2	164.6	212	8.000	5.720
Grimaldi	pN	−5.0	−68.4	210	3.200	1.655
Clavius	N	−58.4	−14.5	210	7.000	4.660

(Continued)

Table 4.3 (Continued)

Name	Period	Lat.	Long.	Diam.	Depth	Fill
Pasteur	pN	−11.6	105.0	210	8.000	5.900
Schwarzschild	N	70.2	120.8	200	4.800	2.900
Bel'kovich	N	61.6	90.2	200	4.000	3.100
Landau	pN	42.0	−119.2	200	6.200	4.200
Galois	pN	−14.5	−152.7	194	7.400	4.800
Schickard	pN	−44.5	−55.1	192	4.800	3.750
Humboldt	UI	−27.0	81.2	186	6.400	5.000
Oppenheimer	N	−35.4	−166.0	180	6.000	4.500
Tsiolkovskiy	UI	−20.6	129.0	180	6.600	4.060
Fabry	pN	42.8	100.8	164	7.000	4.720
Compton North	pN2	61.5	103.8	160	2.600	1.800
Hilbert	N	−18.1	108.3	148	7.300	4.600
Rozdestvenskiy	pN	85.2	−156.2	148	5.400	3.840
Hausen	E	−65.2	−88.5	148	8.600	4.400
Compton	LI	56.0	104.5	138	4.800	3.900
Sklodowska	UI	−18.0	96.1	118	5.800	3.000
Langrenus	E	−8.9	61.2	118	5.440	2.540
Copernicus	C	9.7	−20.1	81	4.800	2.300
Aristoteles	E	50.5	17.0	76	4.400	2.060
Tycho	C	−43.3	−11.3	74	6.000	0.000
King	C	5.0	120.4	68	6.800	4.200

Notes Names of impact features identified after the publication of (Wilhelms 1987) are in italics. Period is from (Wilhelms 1987) except for newer features whose assignment is tentative. Latitude and Longitude are determined from the Kaguya 1/16° DEM. Diameter and Depth are of the apparent crater. Apparent diameter is measured at intercept of the apparent crater with the target surface. Diameters in the literature (Wilhelms 1987) are often measured at elevations part-way up the rims, and are about 15 % larger than apparent diameter. In some cases, the main ring identified by matching the impact model has been identified as inner or outer rings in the literature. Fill depth is measured from the depth at the center of the model of the apparent crater to the level surface of the fill

the residual topographic map that are larger than this, except for the megabasins. Their models will be described in the next chapter. A number of smaller features are included in Table 4.3 to illustrate the Eratosthenian and Copernican periods.

Except for the recently identified features whose names are in italics, all the names listed in Table 4.3 are either IAU names or basins listed by USGS (Wilhelms 1987). Dirichlet-Jackson was named in the Lunar Picture of the Day (Wood 2003). *Lavoisier-Mairan* and *Cardanus-Herodotus* basins, nearly submerged in Oceanus Procellarum, were first identified by radial features in the nearby highlands (Byrne 2004) and their impact nature is clear from the Kaguya DEM. The *Kohlschutter-Leonov* and *Fitzgerald-Jackson* basins (Cook 2002; Oberst 2011) are confirmed here by examination of an intermediate residual DEM. *Compton North,* also identified from an intermediate residual DEM, is an unnamed crater just north of Compton. *Jeans—Priestly* is a crater in the area of Mare Australis.

A few of the probable basins and most of the possible basins (Wilhelms 1987) are not listed in Table 4.3 because they do not have clear topographic signatures that could be matched to the model. Further, they are not necessary to complete the comprehensive topographic model because their circular images do not appear in the residual DEM. That does not establish that they are not basins: for example, the Insularum Basin is revealed by a clear ring of the peaks of its rim, but is so deeply flooded by mare that its depth cannot be estimated by the method described here. Two other large impact features have been identified by circular patterns but do not show the topographic signatures of basins. One is the Procellarum impact feature, suggested by the minerology of rocky outcrops (Nakamura et al. 2012) and the other is the Australe impact feature suggested by a pattern of mare deposits (Byrne 2013). These features may have impacted before the crust fully hardened, absorbing their rims and ejecta fields.

The entry for Grimaldi is for the crater, not the basin. Either the rims of craters surrounding the Grimaldi crater are mistaken for a basin rim or degradation has destroyed the topographic signature of an impact basin larger than the crater.

Some of the basins in Table 4.3 relate to likely impact basins proposed by Frey (2011), specifically Lavoisier-Mairan Basin (TOPO-31), Kohlschutter-Leonov Basin (TOPO-17), Dirichlet-Jackson Basin (TOPO-24), and Fitzgerald-Jackson Basin (TOPO-41). Other features proposed by Frey may be too shallow or degraded to be identified by the methods reported here.

4.14 Summary

Scalable algorithmic models of the apparent crater, rims, and ejecta field have been developed from extensions of the Maxwell-Z model. In addition, models of peak-rings of complex craters and fill by mare or ejecta from later features have been developed. The models have been successfully applied to many large craters and basins; all lunar impact features with clear topographic impact patterns that are over 200 km in apparent diameter. A topographic map of the models of these features is presented in Chap. 5.

Internal rings other than the peak-ring have not been clearly observed other than in the Orientale Basin and the Nectaris Basin, so a formal model has not been developed for them. Although there is some evidence for external rings, it is not clear enough for modeling by this method.

While the parameters of the scalable model have been derived for the self-similar group of large craters and basins described in Chap. 3, they have been successfully used for the many other features of Table 4.3 and have been used (with only the range equations modified for elliptical trajectories) to extrapolate to the megabasins (see Chap. 5).

Chapter 5
The Search for Megabasins to Model the Moon's Large-Scale Topography

5.1 Introduction

An intermediate composite DEM was assembled from models of many impact features and used to produce a residual DEM by subtracting the composite DEM from the current topography: a search for models of megabasins that would improve the fit resulted in finding the NSM, improving the parameters of the SPA, and finding a new smaller megabasin.

Several investigators proposed that impact basins on the near side of the Moon had deposited their ejecta on the far side (Wood 1973), forming the bulge there and other departures from large scale symmetry, as described in Chap. 1. Attempts were made to identify the basin or basins that had created the bulge but further examination of both topography and crustal thickness (Neumann et al. 1996) failed to confirm either the proposed Gargantuan Basin (Cadogan 1974) or the Procellarum Basin (Whittaker 1981) as impact features.

This chapter describes how, with the new model of impact features derived in Chap. 3 and the models of basins and large craters shown in Table 4.3, the search for new megabasins was successful. Two new megabasins, the Near Side Megabasin (NSM) (Byrne 2006, 2007, 2008) and the Chaplygin-Mandel'shtam Basin (CM) were found (Byrne 2012) and new parameters were established for the South Pole-Aitken Basin (SPA).

The tentative name for the CM designates two craters near its edge that have IAU names, Chaplygin and Mandel'shtam, in accordance with USGS tradition. No such designation is possible for the NSM since two features near its edge would be on the far side, and might lead to confusion.

5.2 The Superposition of the Large Impact Features

The model of large impacts described in the previous chapter was applied in custom software to the features listed in Table 4.3 along with the principle of superposition (see Fig. 5.1).

C. J. Byrne, *The Moon's Near Side Megabasin and Far Side Bulge*,
SpringerBriefs in Astronomy, DOI: 10.1007/978-1-4614-6949-0_5,
© Charles J. Byrne 2013

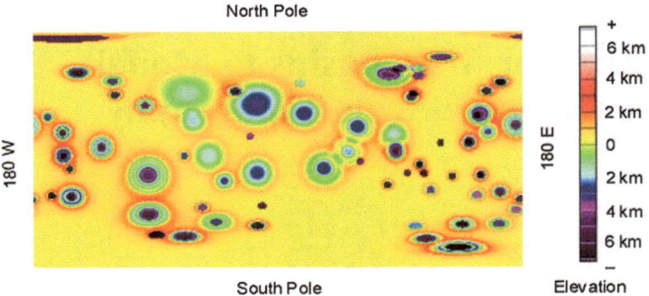

Fig. 5.1 Models of all basins and large craters with apparent diameters of 200 km or greater whose impact nature is confirmed by clear radial elevation profiles. Some selected smaller craters listed in Table 4.3 are also shown here. The SPA is not shown in this figure

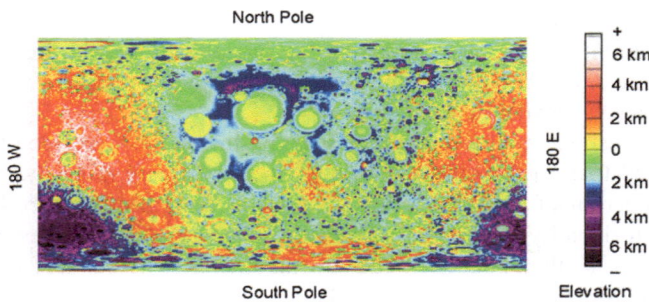

Fig. 5.2 The residual DEM after subtracting the DEM of Fig. 5.1 from the current topography. The large-scale features of the Moon can be seen relatively clearly after the distraction of the large craters and basins are removed. Fill in the crater cavities has been included in the model before subtraction from the current topography

To cover the whole Moon, the Kaguya 1° DEM and the orbital range Eq. 4.21 was used in the software. The SPA is not yet included in this step. As had previously been determined (Petro 2005) the major buildup of ejecta was local; there was no possibility of building up a bulge with ejecta from a potential concentration of basins in this size range. In Fig. 5.1, the apparent craters and their rims and ejecta fields are shown, with peak-rings modeled if observed, but no mare fill is added in this figure. The craters interact according to superposition, so ejecta fill into a crater from nearby impacts of features in this size range is modeled.

As will be done repeatedly in this chapter, the intermediate DEM of Fig. 5.1 is subtracted from the current topography to suggest what other features may be missing. The residual topography after this step is shown in Fig. 5.2.

The intermediate residual DEM of Fig. 5.2 suggests that the major large scale structure of the lunar surface is dominated by the SPA, a far side bulge, and a relatively flat area on the near side of the Moon with a mix of depressions and some mounds.

5.3   The Search Finds the Parameters of the Near Side Megabasin

Following Wood's hypothesis that impacts on the near side may have built the far side bulge with their ejecta, the simulation program was modified to accommodate two larger basins in addition to the list of Table 4.3, with their parameters set through the user interface. In the original search process using the Clementine database, the parameters of the Procellarum Basin and another proposed basin in the southeast of the near side (Feldman et al. 2002) were used. Initially, the SPA was not modeled.

The parameters of the two simulated trial basins were changed by trial and error to minimize the standard deviation of the residual DEM. It was quickly found that the most effective changes were to increase the size of the basins. After a number of increases, the two basins grew to overlap each other. At this point, recognizing that two overlapping basins might represent one very large basin, only one basin was used. Making it larger improved the residual standard deviation even more.

After a minimum standard deviation was reached by varying the parameters of the one large basin, the second simulated basin was used to model the SPA, using parameters from the literature (Garrick-Bethell 2004). Since these parameters included a non-trivial ellipticity for the SPA crater, the program was modified accordingly for both basins.

The model of an elliptical basin allows the calculation of elevation to be varied by azimuth. The apparent diameter becomes an elliptical function of azimuth whose parameters (eccentricity and angle of major axis) are set by the user interface.

The combination of a near side basin with SPA continued to be a better and better fit to the topography as their parameters were adjusted. As the near side basin grew to cover half the Moon and more, its ejecta (that had not escaped the Moon) piled up on the far side until it approached the size of the far side bulge in the current topography. The minimum standard deviation, after adjusting the center latitude and longitude was about half of the standard deviation of the current topography (only about one fourth of the variance).

With minimization of the standard deviation of the residual DEM, the Near Side Megabasin had been found (Byrne 2006, 2007, 2008). The term "megabasin" has been used in other fields to mean a basin so large that it contains other basins within it, a quality that is certainly true of the NSM. This is also true of the SPA, which could have been called the Far Side Megabasin if it did not already have a well-known name. Models of the NSM and the other two megabasins superposed are shown in Fig. 5.3.

5.4   Relocation of the SPA and Another New Megabasin Discovered

When this process was repeated with the Kaguya DEM, it was necessary to revise the parameters of the SPA because of the new data near the South Pole (the Clementine laser altimeter was ineffective beyond 78° North or South latitude

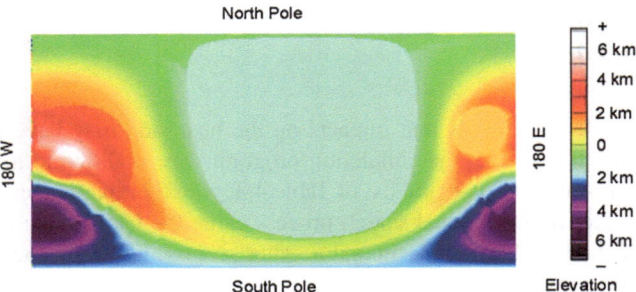

Fig. 5.3 The three megabasins, the Near Side Megabasin (*NSM*), the South Pole-Aitken Basin (*SPA*), and the Chaplygin-Mandel'shtam Basin (*CM*). These models follow the analytic model derived in Chap. 4, with the same parameters that define the target surface. Parameters such as center latitude and longitude, diameters and depth, and fill are assigned to each feature to make a best fit to the current topography of the Moon. Each feature has a *crustal floor* that is probably due to a central melt column that has collapsed. The *far side bulge* is associated with ejecta from the NSM, and has been modified by the SPA

because of the spacecraft's orbital characteristics). However, there was only a small change to the NSM parameters between the Clementine and Kaguya database because most of its topography is within the area of Clementine's accuracy.

Also, with the Kaguya DEM, a third new basin was found from studying the residual DEM after the models of the NSM, SPA, and the list of features from Table 4.3 were subtracted from the topography. This new basin, smaller than SPA but larger than Imbrium, was tentatively designated the Chaplygin-Mandel'shtam Basin (CM) in the USGS convention. It is on the northeast far side. All three giant basins are modeled in Fig. 5.3.

The parameters of the three megabasins, in the same format as Table 4.3, are shown in Table 5.1. The NSM and SPA are actually elliptical, but are represented here as circular features for comparison with the parameters of the features of Table 4.3. The detailed elliptical parameters of the NSM and SPA will be provided in Chaps. 6 and 7.

Figure 5.3 does not show the antinode deposits of the SPA or CM because they do not appear in the residual of Fig. 5.2 or the current topography. Potential reasons for this include:

- The flat floor of the NSM was still molten or plastic when the SPA and CM antinodes were deposited.

Table 5.1 Modeled megabasins

Name	Age	Lat. (°)	Long. (°)	Diam. (km)	Depth (m)	Fill (m)
Near Side Megabasin	pN1	10.0	24.0	6,690	3,950	2,360
South Pole-Aitkin	pN1	−48.0	−171.0	2,684	7,050	700
Chaplygin-Mandel'shtam	pN1	9.6	162.0	1,320	3,800	2,600

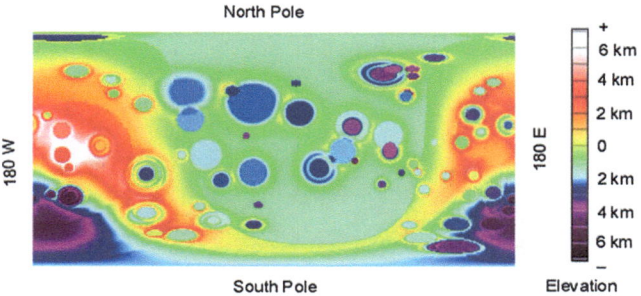

Fig. 5.4   This intermediate composite DEM combines the three megabasins with the impact features of Table 4.3. There is now a stronger resemblance to the current topography. Fill has been included in this model, as in Fig. 5.2

- The SPA and/or CM were prior to the NSM and their antinode deposits were obliterated by the melt column of the NSM.
- The ellipticity of the SPA implied an asymmetric ejection pattern which did not have a large antinode deposit.

Figure 5.4 shows an intermediate composite model with the three megabasins superposed with the impact features of Fig. 5.1. The fill measured from the current topography was added to all the impact features.

Figure 5.5 is the residual DEM corresponding to Fig. 5.4. It shows significant anomalies that do not have the shapes of impact features, suggesting the need for further modeling.

5.5   Additional New Features

Several new medium-sized features can be seen clearly in Fig. 5.5, features that were obscured in the original DEM by the features that have been modeled and removed by subtraction. Most of these features were noted from the Clementine DEM (Byrne 2007) but are much clearer with the Kaguya DEM (Byrne 2012).

5.5.1   Vallis Procellarum

The largest of the new features is an arc-shaped depression running from the Nubium Basin to the Humorum Basin through Oceanus Procellarum and through Mare Frigoris. This depression, which was once thought to be the western edge of the Gigantuan Basin and then the Procellarum Basin, is now seen to be quite a different feature. Since it is associated with the areas of broadest, deepest and

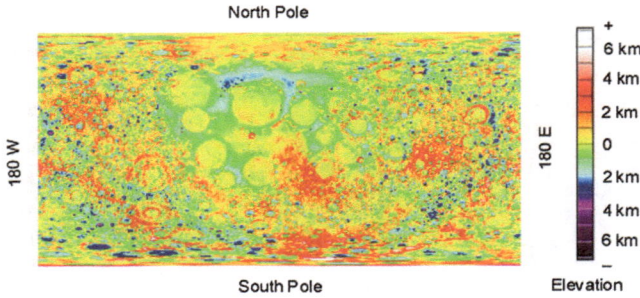

Fig. 5.5 The residual DEM after subtracting the intermediate DEM of Fig. 5.4 from the current topography. Several large features, not showing impact characteristics, appear in this residual DEM

most persistent flows of mare lava, and the region of the largest concentrations of KREEP, it is probably linked to the source of the major lava flows. Its arc is also is associated with the flat floor of the NSM, a point that will be discussed in the next chapter. It is tentatively called Vallis Procellarum.

5.5.2 The Tranquillitatis Depression

A second negative elevation feature is a small circular depression in western Mare Tranquillitatis. This is associated with a large flow of lava that has not only formed Mare Tranquillitatis, but also apparently flowed into the Nectaris Basin, forming Mare Nectaris. Since it is within the area of the NSM melt column, it will be discussed further in the next chapter. It meets the IAU definition of a crater, but if it is an impact crater, it is too heavily flooded to determine its depth by the method used here. There are 4 possible rim peaks exposed over Mare Tranquillitatis that suggest it is an impact feature.

5.5.3 Mons Nectaris in the Central Near Side Highlands

There is clearly a circular mound in the north of the central near side highlands. It has been impacted by the Nectaris Basin event, just as the far side bulge has been impacted by SPA. While this is very clear in the residual DEM of Fig. 5.5, it is also clear in the current topography itself (see Fig. 5.7). This mound could be deposits of material that escaped the Moon in the early phases of ejection from the NSM or the SPA. Such material, having escaped the Moon but not the Earth–Moon system, would mostly have fallen to Earth. However, some could have returned to the Moon (at subsonic velocities, as they left) and formed these mounds. Alternately, this mound could have been formed by uplift caused by plutons from the mantle that did not surface. Whatever its source, this mound is essentially a mountain that is closely associated with Nectaris and might reasonably be called Mons Nectaris to conform to IAU nomenclature.

5.5.4 Three Other Mounds

The other three positive features may or may not be associated with the SPA and the CM (see Chaps. 7 and 8). Otherwise, as in the case of Mons Nectaris, they could be either deposits of material that has escaped the Moon from NSM and/or SPA or they could have been formed by uplift caused by plutons from the mantle that did not rise to the surface.

5.6   The Complete Composite Model

The mounds and depressions of Fig. 5.6 are combined with the megabasins of Fig. 5.3, the large craters and basins of Table 4.3 and Fig. 5.1, and crater fills in a comprehensive model of the Moon in Fig. 5.7, Top. This is a complete model of the current large scale topography of the Moon shown in Fig. 5.7, Bottom and compares very well with that topography.

The residual DEM corresponding to the final model is shown in Fig. 5.8. There are no remaining features obvious in the residual that are larger than 200 km in diameter.

Although there are no obvious features that could be modeled in the residual DEM of Fig. 5.8, there is a subtle pattern that can be perceived. Except for a major region of the central and northern near side, there is a fairly uniform sprinkling of small impact craters just a few pixels across: about 30–100 km in diameter. Yet

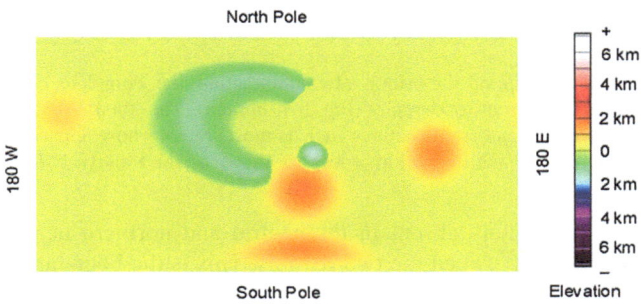

Fig. 5.6 Models of depressions and mounds for features identified in Fig. 5.5

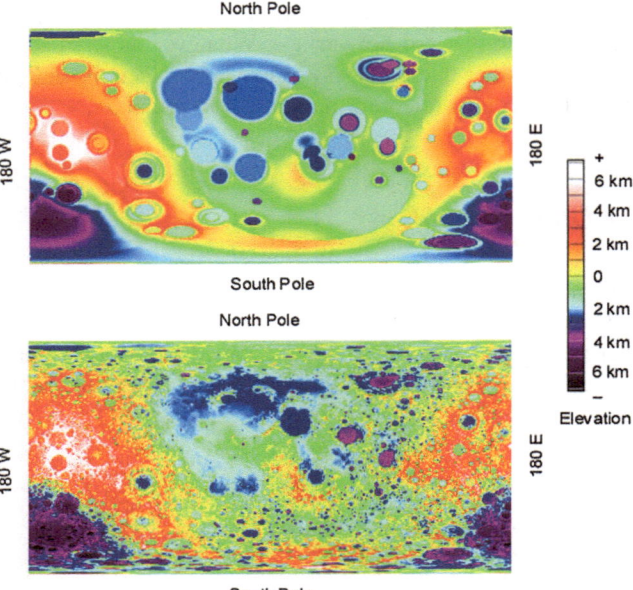

Fig. 5.7 *Top*: The final composite model, combining the megabasins, the impact features of Table 4.3, the mounds and depressions of Fig. 5.5, and fill of the impact features by mare flows, etc. *Bottom*: The current topography of the Moon, Kaguya 1° DEM

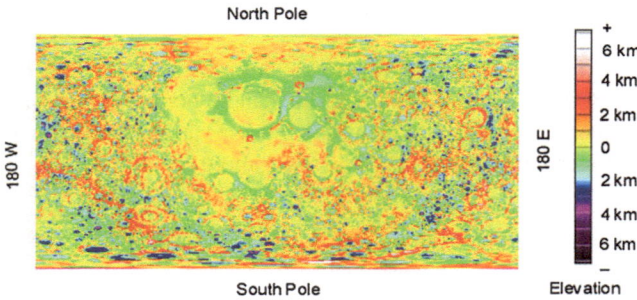

Fig. 5.8 The residual map of the Moon after subtracting the complete composite model (Fig. 5.7, *top*) from the current topography (Fig. 5.7, *bottom*). This final residual map includes imperfections of the model and small features such as most craters whose apparent diameters are less than 200 km (craters as small as 34 km or less are shown on the Kaguya 1° DEM)

they are almost completely absent in the central and northern near side, with a very distinct boundary between the two regions. This is the boundary of the resurfaced area of mare, plains material, and pyroclastic deposits, broad contiguous surfaces. The area is within the western half of the flat floor of the NSM, aligned with

Vallis Procellarum, which forms the western boundary of the resurfaced area. This resurfaced area is not a feature in itself, but a combination of features.

5.7 Discussion of the Complete Composite Model

In solving a picture puzzle, one makes false starts, retreats, and then tries something new until the pieces fit together. Once they fit, there is little doubt that the assemblage is correct. Completing this model of the Moon's topography suggests a similar confidence that all the pieces (at this scale) are found. The nature and cause of some of the pieces is still in doubt but that is for further investigation.

The large scale features of the Moon fall into four categories: the megabasins, the other large craters and basins, the depressions, and the mounds. Most of these features will be found to be intimately identified with the megabasins, especially the Near Side Megabasin. Some further investigations, especially of the megabasins and the new features within them, are discussed in the next three chapters.

Chapter 6
The Near Side Megabasin

6.1 The NSM, the Largest of the Lunar Megabasins

The Near Side Megabasin (NSB) is described in detail, including the shapes of its apparent crater than covered more than half of the Moon and its ejecta field that formed the far side bulge. The three megabasins, NSM, SPA, and CM, interact. They have to: there is not enough Moon to keep them apart. The same condition applied to Earth at this time: it had more surface area but would have attracted proportionately more similar impactors because of its greater gravity. Further, they would have struck with even more kinetic energy than the lunar impactors, creating larger basins. The NSM and the other megabasins prepared the canvas for not only all of the future smaller impacts but also the volcanic outpouring of mare lava. In Chap. 4, the puzzle of the Moon's topography was solved, with the largest and biggest puzzle piece being the NSM, which established the foundation for all the features that followed. If there were earlier giant impacts, they are no longer observable.

This chapter concentrates on the NSM and its properties, examining the crater and its major features in detail and then turning to its rim and ejecta field. As the features are discussed, the explanation of the dichotic nature of the Moon will be revealed. In the last section, the age of the NSM is discussed.

It is difficult for us to perceive a complex structured feature like the NSM when it is wrapped around a sphere and can only be viewed one part at a time. Usually we see a basin all at once, the crater, rim and ejecta field. Although there are ways to produce a map of the surface of a sphere like the Earth or the Moon in a single figure, the required distortion can obscure the geometry of the features. The appropriate projections and their centers need to be carefully chosen to make the shapes clear.

In the case of the NSM, a nearly circular feature whose apparent crater extends over more than half the Moon, a good choice for a map is a pair of Lambert equal-area projections, each with a range of 120°. This projection allows one to look around the east and west limbs and over the poles. One map in Fig. 6.1 is centered on the NSM apparent crater and one on its antinode. Figure 6.1 compares the Kaguya 1° DEM with superimposed models of the NSM and the neighboring SPA and CM megabasins.

C. J. Byrne, *The Moon's Near Side Megabasin and Far Side Bulge*,
SpringerBriefs in Astronomy, DOI: 10.1007/978-1-4614-6949-0_6,

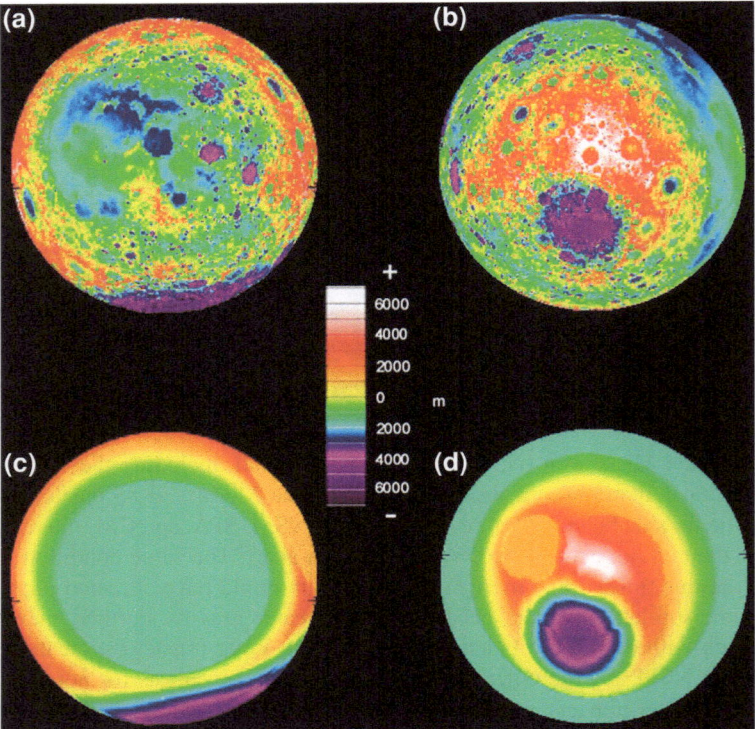

Fig. 6.1 Lambert equal area projections, 120° range, with centers on the NSM apparent cavity (*left*) and its antinode (*right*). The Kaguya 1° DEM is shown in (**a**) and (**b**). Corresponding views of the models of the NSM, SPA, and CM are shown in (**c**) and (**d**)

The NSM has directly influenced the entire surface of the Moon, a large fraction of the lunar crust and a smaller, but significant fraction of the lunar mantle. Figure 6.1 shows that the NSM cavity is slightly elliptic and symmetric around its center. The far side bulge was symmetric about the NSM antinode (with complementary ellipticity) until it was impacted by the SPA. The center of the NSM is at 10°N and 24°E; the antinode is at 10°S and 156°W. The axis between these two points is the center of the dichotic nature of the Moon.

The northern part of the cavity of the SPA has reduced the elevation of the southern part of the far side bulge and SPA ejecta has added to the elevation of the rest of that bulge. The southern rim of the SPA extends over the South Pole onto the southern near side, impacting the southern rim of the NSM. The situation is very similar to interaction between the western rim of the Serenitatis Basin and the eastern rim of the Imbrium Basin; the first impression is that the two rims have destroyed each other where they touch. Actually, the second basin partly ejects the upper part of the rim of the first basin while the rim of the second basin falls into the cavity of the first basin. As a result, in the region of overlapping rims, there is only a little material from either rim left above the original target surface. The SPA is discussed further in Chap. 7.

The CM, described further in Chap. 8, is the smallest megabasin. It has impacted the northwestern far side bulge at 9.6° N and 162° E. Although it is clear in the model, its representation in the DEM of the current topography is more subtle because of the elevation range of the false colors. Its nature is clear in both a radial elevation profile and examination of the residual DEM after all other features are modeled.

The SPA was mapped by the USGS in 1979 (Wilhelms et al. 1979). Why was the NSM not identified long ago, when the Gargantuan and Procellarum Basins were proposed, discussed, and discarded? Three factors have delayed perception of the NSM until 2006 (Byrne 2006):

- The sheer size of the NSM covers more than half the Moon which makes perception of its shape difficult until the axis of symmetry is known.
- Its depth to diameter ratio is only about 0.05 % (the ratio for SPA is about 0.23 %).
- Elevations were not accurately measured over the whole Moon until the Clementine mission produced a DEM.
- Simulations had not advanced until recently to show the remarkable phenomenon of a chaotic melt column formed beneath a megabasin, expanding vertically and then collapsing into a flat floor.

The CM megabasin was identified only after the NSM and SPA megabasins were subtracted from the Kaguya 1° DEM and the residual map examined (Byrne 2012).

The parameters of the NSM and SPA megabasins were determined by trial and error to establish a best-fit to the DEM. A set of 9 parameters was used for each of these two megabasins: latitude, longitude, apparent diameter, eccentricity, inclination of the major axis, apparent depth, fill (elevation of the level crust, measured from the apparent depth), and ejecta launch angle. The equations for the model were as developed in Chap. 4. These parameters were adjusted to achieve a best fit to the Kaguya 1° DEM by minimizing the standard deviation of the residual DEM. The parameters of the CM, being the smallest megabasin, were developed assuming a circular impact feature and were determined from the Kaguya 1/16° DEM using the same methods used for the basins and large craters (see Chap. 3).

The fundamental cause of the lunar dichotomy described in Chap. 1 is the NSM event. A detailed description of the NSM apparent crater, rim, and ejecta field (the far side bulge) follows. The NSM apparent crater will be discussed in Sect. 6.2, the rim and ejecta field in Sect. 6.3, and the age of the NSM in Sect. 6.4. As each NSM feature is described, its relation to the lunar dichotomy is mentioned.

6.2 The Apparent Crater of the NSM

All that we see when we look at the Moon from Earth is within the NSM crater. All of the Ranger impacts and the Surveyor, Luna, and Apollo landings were not only within that crater but also within its flat floor. Landings in the SPA crater and the nearby far side highlands, and samples from there are of great scientific interest but in retrospect, we have concentrated our explorations on the most complex and interesting part of the Moon.

The following subsections describe the range and depth of the NSM crater, the effects of isostatic compensation and crustal thickness, the flat floor and the melt column, the resurfacing by maria, plains and fire fountains, Vallis Procellarum, the relation of the Nectaris Basin to the Near Side Central Highlands, mascons, and the mineral anomalies within the NSM crater.

6.2.1 Range of the NSM

Figure 6.2 shows the Kaguya DEM along with the composite model. The slope of the NSM apparent crater is shaded to show the edge of the flat floor and the edge of the apparent crater, beyond which are the rim and ejecta.

It is very clear from Fig. 6.2 that the NSM apparent crater is nearly circular, but significantly elliptical, indicating a nearly vertical but somewhat oblique impact. Its

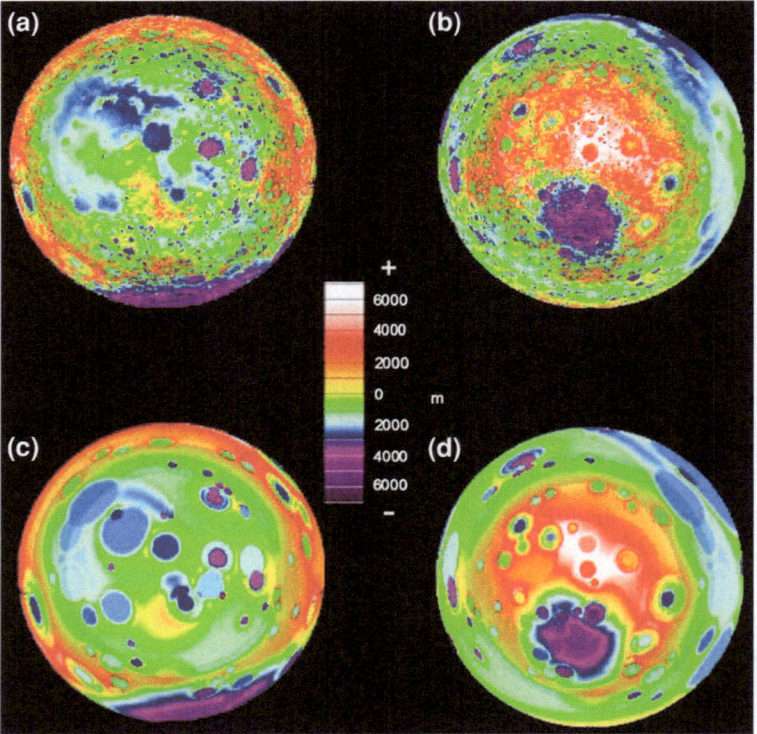

Fig. 6.2 The maps (**a**) and (**b**) are the same as in Fig. 6.1. Maps (**c**) and (**d**) show the comprehensive model. Here, the edges of the NSM flat floor and of its apparent crater are shown by *shading* the area between them. The rim and ejecta field are beyond the outer edge of the *shaded area*. The *two left maps* are centered on the NSM crater and the maps on the *right* are centered on the antinode

center is at 10° N and 24° East. The major axis, rotated 68° counter-clockwise from the north/south direction, is 229° (6,940 km) in diameter and the minor axis is 212° (6,448 km) in diameter. The equivalent diameter (the diameter of a circle of the same area as the ellipse) is 220° (6,690 km). The flat floor, as derived from the model, has the same alignment and eccentricity of its major axis as the apparent crater. Its major axis is 168° (5,097 km) in diameter and its minor axis is 229° (4,736 km) in diameter.

The NSM crater extends over both poles and over the east and west limbs as well. Parts of the slope are visible from Earth, especially in the West. All but the eastern part of the flat floor of the NSM (top of the melt column) is visible from Earth and is responsible for the "Man in the Moon" pattern.

6.2.2 The Depth of the NSM

It would be nice to have an elevation traverse of the NSM, but it would be distorted by the other large features of the Moon. By averaging elevation over azimuth, the distortion could be minimized but the SPA megabasin would still cause considerable distortion. Also, the Korolev Basin, that has the antinode of the NSM within its crater, would distort the center of the NSM ejecta field. To avoid these problems, models of the SPA, the Korolev Basin, and the nearby Gallois crater were subtracted from the Kaguya DEM to produce the radial elevation profile shown in Fig. 6.3.

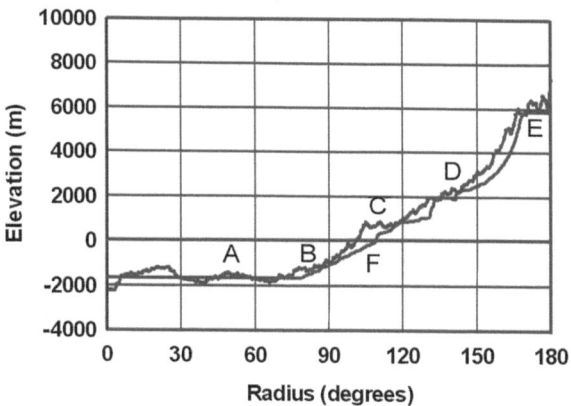

Fig. 6.3 *Rough line* a radial elevation profile of the Moon, from the center of the NSM apparent crater. *Smooth line* model of the NSM. Both are circular approximations to a somewhat elliptical feature. *A* the flat floor formed by the collapse of the melt column. *B* the slope of the apparent crater. *C* the rim. *D* the ejecta field, piled high because the landing zones have a smaller circumference than the ejecta zones. *E* the peak of the far side bulge, where ejecta lands from all directions with a high velocity. *F* the edge of the apparent crater, where it intersects the original target surface (set at 0° elevation in this graph)

6.2.3 Isostatic Compensation and Crustal Thickness

The NSM occurred very early in the history of the Moon, when the crust, newly formed from the magma ocean, had recently hardened. The free-air gravity map in Fig. 7.6 shows no anomalies of the scale of the NSM, indicating full isostatic compensation of not only the crater with its flat floor but also the rim and ejecta field. That is, the mantle must have risen to compensate for the crater and been depressed below the far side bulge (Neumann et al. 1996; Wieczorek and Phillips 1998; Wieczorek et al. 1999).

Assumptions about the density of the crust (2.80 g/cm³) and mantle (3.36 g/cm³) (Hikida and Wieczorek 2007) imply that complete isostatic compensation would reduce initial topography to current topography by a factor of 6.0. The reasoning that led to the density assumptions above can be found in (Hikida and Mizutani 2005). The Airy model of isostatic compensation assumes that all motion in the crust is vertical; movement is through fracturing. Then the shape of a surface feature is preserved except for the proportional reduction in elevation variations.

The Airy model relates the factor of equilibrium compensation to densities of the crust and mantle as follows:

$$z_{iso} = z \cdot (\rho_{mantle} - \rho_{crust})/\rho_{mantle} = z/6.0 \qquad (6.1)$$

where z is an initial elevation variation and z_{iso} is the variation after equilibrium compensation.

Recent analysis of GRAIL data has indicated a porosity of highland crust of at least 12%. This would imply of compensation of 3.75. This has not been included in the folllowing discussion.

The parameters of the model of the NSM have been adjusted to the current shape of the Moon. As a result of normalizing the apparent cavity by apparent depth and apparent diameter independently to preserve the observed self-similarity of impact features, application of the model to a compensated feature absorbs not only complete but also partial Airy isostatic compensation. The close agreement between the model of the NSM and the current topography establishes that the assumption that the Airy model is appropriate for the early Moon as opposed to the Vening Meinesz flexural model which allows for some horizontal distribution of the strain.

The current apparent depth of the NSM crater, estimated from the exposed slope and the negative cosine model, is 3,950 km. Since the free air gravity shows that the NSM is at equilibrium compensation, the initial apparent depth of the NSM crater (short lived because of the immediate expansion and collapse of the melt column) was 6 times 3,950 km, or 23.7 km. For comparison, the Challenger Deep in the Pacific Ocean is 11,030 meters below sea level. The depth of the NSM melt column itself would be much greater, extending beyond the crust far into the mantle to a depth of perhaps 500 or 600 km.

The apparent depth D_a of the NSM can be combined with the equivalent apparent radius R_a (6,690 km/2 = 3,345 km) and the cosine model of the apparent crater to estimate the volume ejected from the apparent crater. That volume is a special case of the volume of flat fill, such as the volume of mare fill in a basin. The calculation can be done by integrating cylinders of radius r, height D_a Cos(r/ R_a), and thickness dr. The range of r would be from r = 0 to the radius where the fill level intersects the cosine. The equation for the volume of fill V_f is:

$$V_f = D_a \cdot R_a{}^2 \cdot \int_0^{\frac{2}{\pi}A \cos(1-F/D_a)} 2 \cdot \pi \cdot \frac{r}{R_a} \cdot Cos\left(\frac{r}{R_a} \cdot \frac{\pi}{2}\right) \cdot \frac{1}{R_a} dr \quad (6.2)$$

where D_a is the apparent depth, F is the fill depth, and R_a is the effective apparent radius, the square root of the product of the major and minor apparent radii for elliptical craters.

Simplifying the integral,

$$V_f = 2 \cdot \pi \cdot D_a \cdot R_a{}^2 \cdot \int_0^{\frac{2}{\pi} \cdot A \cos(1-F')} r' \cdot Cos\left(r' \cdot \frac{\pi}{2}\right) \cdot dr' \quad (6.3)$$

where $r' = r \cdot /R_a$ and $F' = F/D_a$.

After integration and applying the limits

$$V_f = \frac{8}{\pi} \cdot D_a \cdot R_a{}^2 \cdot \left[-F' + A\cos\left(1 - F'\right) \cdot \left(1 - \left(1 - F'\right)^2\right)^{0.5} \right] \quad (6.4)$$

To find the total volume ejected from the NSM (the volume of the initial apparent cavity), set F = 1;

$$V_e = \frac{8}{\pi} \cdot \left(\frac{\pi}{2} - 1\right) \cdot D_a \cdot R_a{}^2 \quad (6.5)$$

$$V_e \approx 1.454 \cdot 23.7 \, \text{km} \cdot (3,345 \, \text{km})^2 \approx 386 \cdot 10^6 \, \text{km}^3 \quad (6.6)$$

The volume of 386 million km^3 seems like a large number but it is not large in comparison with the Moon's total volume of 21,958 million km^3. Thus NSM ejected about 1.8 % of the Moon's total volume, while its crater covered more than half of its surface and its rim and ejecta field covered the rest. A far greater volume was subjected to violent phase changes in the melt column below the apparent crater.

Figure 6.4 shows the cumulative volume ejected from the NSM as a function of normalized internal radius. While 336 million km^3 escaped from the Moon, the remaining 50 million km^3 was deposited and formed the far side bulge as described in Sect. 6.3.

The material that escaped the Moon stayed in the Earth-Moon system. Most of that has ultimately come to the Earth, but some may have returned to the Moon, possibly forming one or more of the mounds described in Chap. 4.

Fig. 6.4 The cumulative volume of crust ejected from the NSM as the ejection cone progressed from the center to the edge of the apparent crater. Up to a normalized radius of 0.79, all of the material escaped from the Moon. Beyond that radius, the material fell back to the Moon, forming the far side bulge

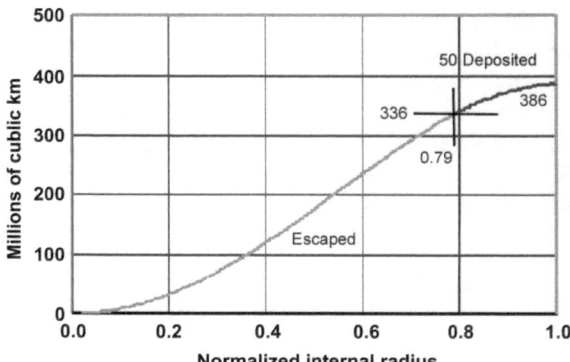

Fig. 6.5 The NSM struck an apparently uniform layer of crust, piling a mound of ejecta at its antipode. The surface elevations here are prior to isostatic compensation. The melt column is not shown

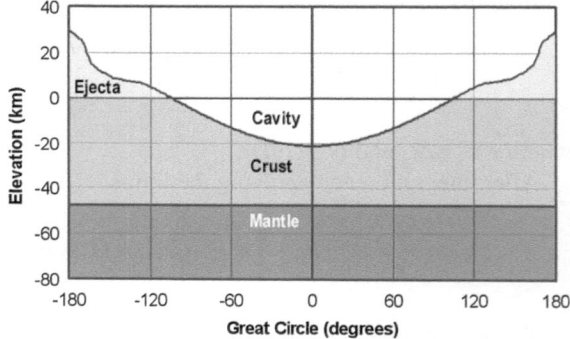

Now that we have a feel for the vertical scale of the early pre-compensation events, we can examine the interplay between two of the megabasins. The cross section of the crust and mantle after the NSM event and before isostatic compensation is shown in Fig. 6.5. The melt column is not shown here, although it was actually part of the event.

Clearly, this single event had a profound effect on the entire crust, establishing the major lunar elevation dichotomy. In following subsections, the history of additional events will be described that follow the NSM and are strongly dependent on the dichotomy it established for their expression.

Another impact formed the South Pole-Aitken Basin (Fig. 6.6). This profile is taken along the great circle that connects the centers of the two giant basins and shows the interaction between the two basins. The SPA depth is much greater than the NSM depth, probably because it encountered the fragmented, porous NSM ejecta while the NSM encountered newly solidified crust.

After equilibrium compensation took place (Fig. 6.7), the mantle rose below the craters and became depressed below the ejecta. Consequently, the surface elevation variations were reduced by a factor of 6.

Fig. 6.6 This profile is along the great circle that connects the centers of the two giant basins and their antipodes. It shows the superposition of the SPA and the NSM. The elevation variation is as it would have been if neither NSM nor SPA had been compensated

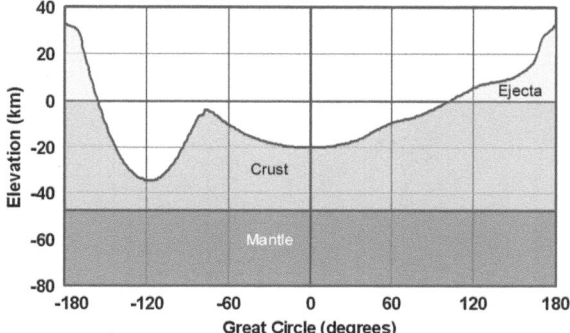

Fig. 6.7 In time, the mantle pushed upward to compress the ejection cavities and the piles of ejecta pushed the mantle downward, leaving a surface variation only one-sixth of the initial variation

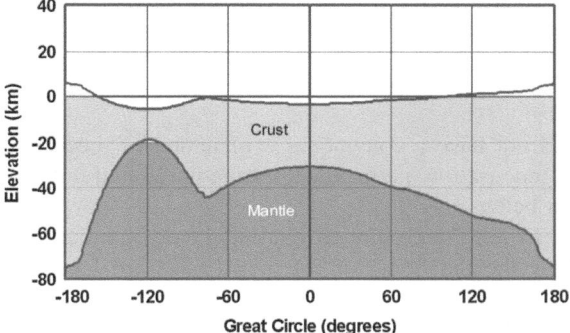

Figure 6.8 shows the current topography along the great circle at a larger scale. It also shows the flat floor of the subsided melt columns, which actually formed before isostatic compensation of each of the megabasins. Initially the depth of the flat crust relative to the apparent crater was probably higher than it is in the current topography because the material was molten and expanded. As the material cooled and recrystallized, it dropped to its current relative depth and underwent further isostatic compensation.

6.2.4 Description of the Flat Floor and Melt Column of the NSM

The idea of a crustal flat floor in the megabasins was difficult to accept. The flat floors of the familiar near side basins were formed by dark basaltic flows of low viscosity lava. The floor of the NSM, where exposed in the near side central highlands and in rims of the near side basins is clearly crust, possibly regenerated from a mix of mantle and crust in the melt column. In my earlier work on the NSM, I

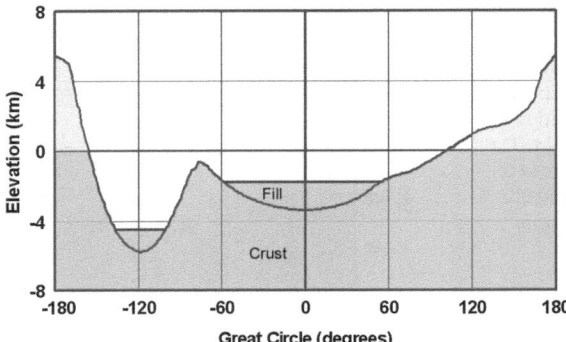

Fig. 6.8 This scaled-up figure shows the current topography of the Moon. The flat floor resulting from the melt column of each megabasin formed before isostatic compensation. Note that the rims of these two megabasins have mutually eliminated each other where they overlap near the South Pole. This is directly observable in the DEM as well

did not have a convincing explanation of the flat floor, attributing it to isotropic compensation, perhaps with some horizontal flow. The melt column explanation is far better.

Recent three-dimensional dynamic simulations of the SPA event illustrate how the flat floors of the megabasins are formed (Ivanov 2007; Stewart 2011). Examples of these simulations will be discussed in Chap. 7. The other megabasins show the same characteristics, but have not yet been simulated.

For megabasins, a melt column extends far below the apparent crater. Initially, the melt column is a chaotic mixture of fractured, melted, and vaporized material extending down through the crust and well into the mantle. The simulations show that the diameter of the melt column would have been equal to the diameter of the flat floor. Within melt columns, the turbulence causes mantle material to be mixed with crustal material and components of the incompatible layer. This has had an influence on the later effusion of lava onto the surface of the NSM crater and the distribution of element and mineral anomalies there.

While some mantle material may have come to the surface, the column of melt would have re-crystalized in a repeat of the solidification of the magma ocean. The lighter crystals of anorthosite would float to the surface, reforming that portion of the crust that had not been ejected. The mantle material returned to the lower part of the column and the incompatible material (KREEP) would have been between the two layers, possibly concentrated by the second separation by re-crystallization and by horizontal turbulence in the melt column.

The diameter of the melt column, according to the simulations, is the diameter of the flat floor, 162° (4,913 km), nearly half the Moon. The depth of the NSM melt column could be estimated by new simulations. A layer of seismic activity was measured by Apollo seismographs, about 500 km below the surface. It has been proposed that that may have marked the lower boundary of the magma ocean. However, it could also mark the lower boundary of the melt column. A modest seismometer network on the far side would distinguish these two suggestions and perhaps find activity below the SPA. In any case, the total volume of the NSM melt column would be a significant fraction of the volume of the early magma ocean. It would have taken a very long time to cool, less than but comparable to the time the magma ocean itself needed to solidify. That interval has been estimated to be as short as 10 million years to as long as 200 million years by various authors (Shearer et al. 2006).

The NSM melt column is a major contributor to the dichotic history of the Moon; it is a massive feature of the near side while the SPA and CM are much smaller features on the near side that are much less complex.

6.2.5 Resurfacing in the Crater of the NSM: Maria, Plains, and Fire Fountains

The greater part of the crater of the NSM, especially in the flat floor, has been deeply resurfaced. Careful examination of the elevation map (Fig. 6.9, left) shows that most of the surface is speckled with craters in the range of 30–90 km. These presumably resulted from bombardment over the whole moon, but are noticeably absent in a large region within the NSM crater. The albedo map shows that the dark maria have obscured the craters in most, but not all of that resurfaced region. Additional areas have been resurfaced but are outside of the maria.

Nearly all the maria on the Moon are within the NSM crater, either on the flat floor or the adjacent slope. Of the 10 largest basins on the Moon (not counting the megabasins: see Table 4.3) all but the Orientale Basin are within the flat floor of the NSM and all have maria within their craters. The depth of this concentration of lava flows is deepest where the large basins were flooded and often shallow between them, where it flowed over the level crust that was formed after the collapse of the NSM melt column. In general, lava flows on the Moon are very flat, indicating a low viscosity, more like water than terrestrial lava flows. The reason

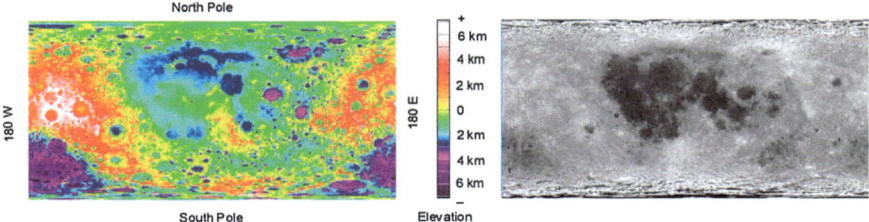

Fig. 6.9 *Left* the Kaguya 1° DEM. *Right* the albedo map of the Moon. The elevation map shows a fairly uniform pattern of small craters that generally correlate with the *dark areas* of maria shown in the albedo map. Source of the albedo map USGS, Map-a-Planet, Clementine albedo

for this has been attributed to much higher concentrations of volatiles on Earth, which expand into vesicles within the lava as it reaches the surface. The resulting foamy consistency of the terrestrial lava increases its viscosity. Despite the flatness of each flow, the areas of deepest lava flows within basins often show a depressed surface elevation. This is probably due to a combination of factors: contraction of the lava as it hardens, depression of the supporting crust by the addition of the weight of the lava, and subsidence of the crust due to depletion of the subsurface source of the lava. There is also evidence of "pumping" of the lava flows, as often happens in calderas on Earth. One example is a shoreline raised above the current level of Mare Nubium. Evidence of subsidence of the crust below northern Mare Imbrium is discussed in the following section.

In addition to mare flooding, there are other kinds of resurfacing that have obscured smaller craters. The ejecta field of a large impact feature can obscure smaller neighboring craters, especially the ejecta blanket which extends to about twice the apparent radius and consists of radial ridges and troughs, boulders and fractured rubble. The Imbrium Basin and the Orientale Basin are examples of impact features whose ejecta fields can be seen to obscure smaller craters. The light albedo area on the near side of the North Pole has been covered with ejecta from the Imbrium Basin, evidence that the heavy bombardment was nearly complete at the time of that basin, about 3.9 Ga ago. Older large basins and the megabasins have also obscured small craters but a continuing high rate of bombardment replaced them.

Beyond a basin's ejecta blanket is often a light plains unit composed of relatively smooth powdered material that can also cover craters. There are also dark plains units consisting of pyroclastic deposits of glass beads erupted from fire fountains. These units are typically near the boundaries of maria where lava flows approach highlands. Many such beads from Apollo samples have recently been found to have porous rinds that contain samples of the propellant gas, typically carbon dioxide, fluorine, and sulfur.

6.2.6 Non-Impact Features: The Vallis Procellarum Melt Pool

Figure 6.10 shows the residual DEM before and after the 6 models of the non-impact features were added (Fig. 5.6). Of the six features, vallis procellarum is related to the NSM melt column. The Mons Nectaris and the Tranquillitatis depression are related to the Nectaris Basin and are discussed in Sect. 6.2.7. The Northern Far Side Mound and the Eastern Limb Mound are related to the NSM ejecta field and are discussed in Sect. 6.3. The Southern Near Side Mound is related to the intersection of the NSM and SPA and is discussed in Chap. 7.

The arc of Vallis Procellarum is entirely covered with mare lava flows (compare Fig. 6.9, right). Various lines of evidence suggest that the lava flow in this large region is due to a major pool of mantle material that has been melted by radioactivity from the incompatible materials concentrated in (and probably by) the NSM

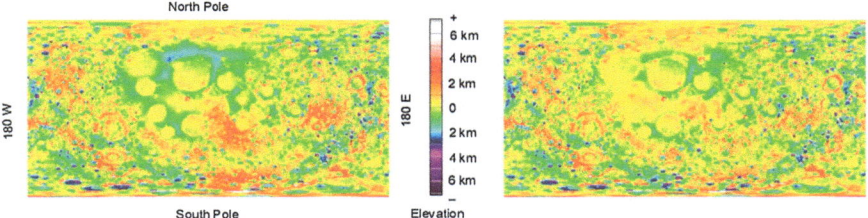

Fig. 6.10 The residual DEM on the *left* was computed with all models except those of Vallis Procellarum, the Tranquillitatis depression, the Mons Nectaris, the Southern Near Side Mound, the Eastern Limb Mound, and the Northern Far Side Mound. The comprehensive residual model DEM is on the *right*. The models of these depressions and mounds are shown in Fig. 5.5

melt column. The Vallis Procellarum was the anchor of the proposed Gargantuan and Procellarum Basins, but further data on both topography and crustal thickness disproved the existence of such an impact basin. This was confirmed by identification of the much larger NSM whose crater encloses Vallis Procellarum.

Vallis Procellarum has within it three drowned basins which do not constitute the depression, but were probably formed by earlier independent events. These basins are circular features with nearly drowned arcs of rims and some radial troughs and ridges (Byrne 2004) that are typical of impact basins. They are formed independently of the large arc-shaped feature Vallis Procellarum. Of these three basins, the Flamsteed-Billy Basin, is one of the basins identified by the USGS (Wilhelms 1987). The Lavoisier-Mairan Basin was also identified by its thorium anomaly by other authors. The Cardanus-Herodotus Basin lies between the other two.

The model of Vallis Procellarum is based on a circular arc whose center is on the major axis of the NSM but is offset to the northwest of its center by 0.25 of the major radius. The arc extends from its center over an azimuth range of 90° on either side of the NSM major axis. Its normalized width along the NSM major axis is from an inner limit of 0.21 to an outer limit of 0.75 (the fractions are normalized on the major radius). The depth, a maximum of 1,250 m, is constant with azimuth and its radial function is sinusoidal for an angle from 0 to π as the radius runs from the inner limit to its outer limit.

Although the model sounds complicated, it is a reasonably smooth function. The model is not based on any theory of formation but is simply a good empirical fit for the new feature needed to complete the comprehensive model of the lunar topography. The parameters of the model; maximum depth, offset from the NSM center, and the inner and outer radii, were set to minimize the standard deviation of the residual DEM. Again, see Fig. 5.6 for a DEM of the resulting model of Vallis Procellarum.

What could be the explanation of Vallis Procellarum? It is in the area of the largest volume of basaltic lava on the Moon. It is located nearly all within the flat floor of the NSM but partly on the adjacent slope. Since it is symmetrical about the major axis of the somewhat elliptical NSM, it could be on that edge of the melt column that is downstream of the impact direction, where the turbulence within the melt column might be greatest. The possible uplift and purification of the incompatible elements by re-crystallization may well have combined, through radioactive heating, to formation of a pool of melted mantle material, the source of lava plumes. Of course, these thoughts are speculative until further physical simulations are run. Data from the GRAIL mission may also reveal the depth and nature of the melt pool.

Whatever the early cause of the great outpouring of basalt in this area, melted over time by intense radioactivity, the deep flows subsequently cooled and then contracted, resulting in part of the depression. The outpouring would of course have depleted the melt pool. The crust in the area was both pulled downward by the cooling and contracting subsurface melt pool and pushed downward by the heavy burden of erupted lava, causing collapse of the crust. That subsidence of the crust would have accounted for the rest of the Vallis Procellarum depression.

Evidence of such subsidence is especially clear in the Imbrium Basin. The northern half of that Mare Imbrium slopes sharply to the North, into Vallis Procellarum. In no other mare is a slope of that nature to be found. Further, the northern rim of Mare Imbrium is much lower than the southern rim. A reasonable interpretation of that slope and rim tilting is that the mare was emplaced and fully hardened before the cooling of the subsurface melt pool and the northern portion of the Imbrium Basin dropped to replace the depleted volume of the melt pool. One way of looking at the process is that as the melt pool produced eruptions through the crust to the surface, the crust also sunk into the melt pool.

There is an irony to this proposal. After decades of discussion of whether the lunar craters were due to impact and volcanism, it turns out that the volcanism that flooded the maria in the NSM was itself facilitated by an impact event.

6.2.7 The Relation of the Nectaris Basin to Mons Nectaris and the Tranquillitatis Depression

The Nectaris Basin has a very complex relationship with Mons Nectaris and the Tranquillitatis depression. The elevation map and photographic coverage of the area are shown in Fig. 6.11.

Analysis of the interplay of these three features illustrates the value of modeling to deconstruct the history of the Moon. It is very clear from the Kaguya 1° DEM (Fig. 6.11, left) that Mons Nectaris was an approximately circular feature before it was impacted by the Nectaris Basin event; much like the far side bulge was impacted by SPA. The center of Mons Nectaris is at 26°S and 20°E and its diameter is 2,120 km. Its cross section is a raised cosine (range $-\pi$ to $+\pi$) with a height of 1,900 m.

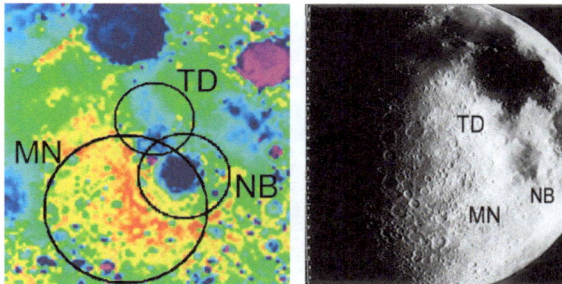

Fig. 6.11 The map on the *left* is a segment of the Kaguya 1° DEM for the region of the Nectaris Basin (*NB*), including Mons Nectaris (*MN*) and the Tranquillitatis depression (*TD*). The photograph on the *right* includes the features of the map. *Source* NASA, Lunar Orbiter LO4-186 M, LPI, (Byrne 2005)

The Tranquillitatis depression was found by searching through a preliminary residual composite DEM for circular negative features. Although several circular features were found and characterized as new impact craters or basins, this was the only negative circular feature examined that did not show the signature of an impact: it lacked both rim and ejecta blanket. The center of the Tranquillitatis depression is at 1°S and 27.1°E and its diameter is 1,360 km. Its cross section is a negative raised cosine (range $-\pi/2$ to $+\pi/2$) with a depth of 1,950 m.

A possible explanation of these observations is that a source of basaltic lava rose at the location of the Tranquillitatis depression and flooded the surrounding area of Sinus Asperitatis until it encountered the rim of the Nectaris Basin and overflowed a low part of the rim. A resulting flow of lava destroyed an arc of the rim and filled the Nectaris Basin. The flow formed a trough between the Tranquillitatis depression and Mare Nectaris. The trough entered the Nectaris Basin at the foot of Mons Nectaris, where it would have been lower than the rim to the west.

After the flow subsided, the trough and the southern part of Sinus Asperitatis emptied, exposing surface roughness. The lava at the source also subsided because of either loss of pressure in the plume or contraction on hardening (Fig. 6.12).

6.2.8 Mascons

Mascon is "short" for a mass concentration (Muller and Sjogren 1968). Its effect is to produce a positive anomaly in the gravity field. There are two related definitions, one from the viewpoint of spacecraft navigation and the other from understanding the nature and history of the lunar surface. The navigation definition concerns those mascons that are strong enough to perturb the flight of a spacecraft sufficiently to affect its mission goals. Lunar scientists are interested in the physical implications of a mascon; what is the nature of the concentrated mass? Where does it come

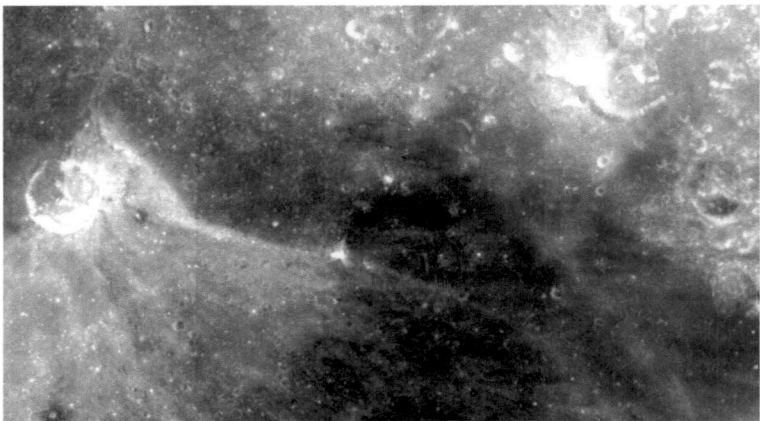

Fig. 6.12   This is the trough that directed lava from Sinus Asperitatis into Mare Nectaris. *Source* USGS Map-a-Planet, Clementine Albedo

from? What does it imply about the surface environment? A mascon can be either a mountain of the same material of its surroundings or a concentration of relatively high density material. In either case, it has not achieved isostatic equilibrium with the underlying mantle; instead, its weight is supported by stress in the crust.

Large lunar mascons in the crater of the NSM (see Fig. 1.6) are associated with the maria in the Imbrium, Serenitatis, Crisium, and Nectaris basins. Most large mascons are in the flat crust of the NSM where basaltic lava is plentiful, but smaller mascons are of considerable scientific interest. The basin itself may have achieved isostatic equilibrium when the crust and mantle were relatively plastic but the maria, deposited much later, is supported by a strong well-hardened crust and therefore acts like a point source of an anomalous free-air gravity field. From a spacecraft view, the closer the orbit is to the surface, the stronger the anomaly. From a science view, the strength of the anomaly can give insight into the difference of density between the mascon material and the crust, the volume of the material, and the bearing strength of the crust.

An estimate of the volume of the dense basaltic lava in a basin may be calculated from the model of a cavity and the level of fill. The resultant volumes for the four largest mascons of Fig. 1.6 have been calculated from the parameters of Table 4.3 and Eq. (6.4). Elevation of the level of fill and the depth to the center of gravity of the fill are also shown in Table 6.1.

Some maria have within them the source of the lava flow, the pluton from the mantle, the equivalent of a volcanic pipe on Earth. Residual lava in the pipe may add to the effective volume of the mascon. Some flooded basins may have internal rings, which would reduce the volume of the mascon. These considerations could either increase or decrease the volume estimates of Table 6.1. It is hoped that this data, along with an estimate of the mare density, would permit the correlation of these equivalent point sources with the observed gravity anomalies. In any case, it is now clear that the dichotic distribution of the largest mascons is associated with the NSM.

Table 6.1 Mascon volumes

Mare	Volume, 10^3 km^3	Fill elevation, km	Depth to C. G., m
Imbrium	1,499	−2.36	1.11
Serenitatis	594	−2.60	0.82
Nectaris	306	−2.74	0.08
Crisium	290	−3.58	0.12

The history of the relation of mascons with spacecraft navigation is interesting. It is difficult to say when the mascons were discovered. The first indication that something new occurred was when a Ranger spacecraft arrived at the Moon about 2 s later than predicted. However, there were many sources of error, so although mascons might have contributed, they were not identified at the time. Luna 10 orbited the Moon in April and May, 1966 and the Soviets reported "… a highly distorted gravity field, suggesting a non-uniform mass distribution". (National Space Science Data Center).

A sequence that led to characterization of mascons was related recently by Matt Grogan, a member of Boeing's Lunar Orbiter operations team who later joined the Apollo navigation team (Grogan 2006). The sequence started with Lunar Orbiter 1 after it went into low orbit in August of 1966. A Moon gravity model had been chosen for Lunar Orbiter and for Apollo sometime between 1964 and 1966, based on spherical harmonics. During the first two missions, the JPL Deep Space Network tracking data showed residual Doppler errors that exceeded expectations. As a result, the spacecraft was not exactly in the right position when the camera took the photographs. During the Lunar Orbiter 3 mission, these errors, usually downrange, were found to be in a pattern, allowing them to be predicted and corrected by a process termed "Kentucky windage": the expected error was corrected by changing the time of the exposure.

Douglas Lloyd of Bellcomm represented the Apollo program at Lunar Orbiter operations. He reported the downrange tracking problem and its implications to the Apollo landing accuracy. My response, in planning the program for selection and certification for Apollo landing sites, was to ask the landing site analyst at NASA Manned Flight Center to double the area vetted for safety.

In 1968, researchers at JPL (Muller and Sjogren 1968) published the word "mascons" and established a 1:1 correlation between the large mascons and ringed basins. By this time, mascons were the subject of an Apollo "Tiger Team".

After the Lunar Orbiter program, Matt Grogan joined NASA to work on the Apollo navigation team and carry his approach to correcting expected navigation errors. After the Lunar Orbiter program, Apollo adopted a new gravity model with 5 spherical harmonics but there still had not been a low orbit polar mission with far side monitoring (which would not be complete until Kaguya) so the mascons were relatively understood but not yet completely known.

The new Apollo approach was to monitor the Doppler data during the initial orbits around the Moon, supplemented by range data. Errors relative to prediction would be identified and used to make predictions for future orbital positions, extracting systematic errors. This process was implemented and tested during the

Apollo 10 lunar orbital mission. The intent for precision landing missions was to modify the Lunar Module guidance computer to allow a manual input from the pilot to enter a correction at 2 min after Lunar Descent Initiation (when the Lunar Module thruster was fired). This modification would not be ready until Apollo 12. The Apollo 11 would land on a very large smooth area in Mare Tranquillitatis. It landed safely, 8 km downstream from its intended target, but within the allowance for redirection by Conrad to find a suitable landing spot.

The plan that was derived from "Kentucky Windage" was successful for all missions from Apollo 12 on. The touchdown for Apollo 12 was 165 m away from its target point in the Surveyor crater, well within the landing precision that was inherent in the program once the mascon problem was solved.

So when were mascons discovered? You be the judge.

6.2.9 Mineral Anomalies

The mineral anomalies tell the story of chaotic turbulence in the melt columns of megabasins. They are associated with variations in the source and history of the mix of mantle and crust in the expanding melt column, its subsidence, and its subsequent history. Simulations of SPA will be presented in Chap. 7: none are available for the NSM as yet. The surface manifestations of the NSM will be discussed here.

While there are smaller scale variations, the overall pattern of mineral anomalies (Sect. 1.5), especially their strongest concentrations which are in the NSM, is clear. Iron, titanium, and thorium follow the pattern of the maria along the major axis of the NSM, to the west of its center toward Vallis Pocelllarum. The KREEP elements of the incompatible layer are especially strong there. The pattern is symmetrical about the major axis. A weaker pattern of the same anomalies holds in the east of the center, but it follows the pattern of large basins northeast of the major axis; few basins are southeast of the center.

It is clear that the propensity for mare flow is along the major axis of the NSM. A possible interpretation is that the major axis indicates the direction of a modestly oblique impact, perhaps 40° from the vertical approaching from the southeast toward the northwest. In the melt column, there may have been a concentration of the incompatible layer that contains a relatively high concentration of uranium and other radioactive elements, especially to the downstream side of the impact point. Perhaps both the radioactive elements and the anomalies of element concentrations were refined by melting and recrystallization, as is used in the refining of semiconductor materials.

In time, up to hundreds of million years after the Imbrium impact, estimated to be at 3.9 Ga, the concentrated radioactivity heated pools of melted basalt from the mantle. The melt rose as plutons toward the surface, preferentially where the crust was thinnest due to basin formation. The plutons carried the element concentration from their sources, perhaps picking up additional minerals from the layers they passed.

Other, weaker anomalies are present in the floor of the SPA, strengthening the association of these variations in element abundance with the melt columns of the larger megabasins.

6.3 The Rim and Ejecta Field of the NSM: The Far Side Bulge

The NSM rim rises from the slope of the NSM apparent crater. The rim and the adjacent ejecta field form the far side bulge. Like the NSM apparent crater, the far side bulge has undergone complete isostatic compensation and has deep roots to a depth of six times the current bulge, depressing the mantle beneath it and the original crust as well. Together with the current bulge, this accounts for the deepening of the far side crust. The radial elevation profile of the rim and ejecta field along with the profile of the model can be seen in Fig. 6.3.

6.3.1 The NSM Rim

The NSM rim has been so eroded and modified by later bombardment that it is difficult to see at most locations, although it is clear in the radial profile (Fig. 6.3). The rim runs through two reasonably young (Imbrian period) features; the crater Tsiolkovsky (see Fig. 6.13) and the Orientale Basin (see Fig. 2.7). The shape of the rim (the target surface for the Orientale Basin) is clearly delineated in the

Fig. 6.13 The rim of the NSM has been directly impacted by the crater Tsiolkovsky. Note the horizontal teraces on the northern rim of Tsiolkovsky, and the unconsolidated material rising above them toward the east. *Source* NASA, Lunar Orbiter LO3-121H1, LPI, (Byrne 2008)

radial elevation profiles by quadrant in Fig. 3.1 and the elevation averaged by azimuth in Fig. 3.2.

The rim drops off slightly beyond its peak unlike most rims of impact features, which drop off at a greater slope. The reason is that as the radius beyond the peak increases, the ejected material that is deposited in each ring of incremental radius decreases but the circumference of that ring decreases more. This condition progresses until the antipode is reached. There, the incoming ejecta converges from all sides, at an angle of about 45°. There is a catastrophic collision and the ejecta in that region is spread over a large area.

6.3.2 The Layered Structure of the NSM Ejecta Field

The scaled model of a hypervelocity impact establishes how the NSM produced two distinct layers of crustal ejecta between the primitive crust and the ejecta from subsequent impacts. The sequence of the deposit of the two layers is shown in Fig. 6.14.

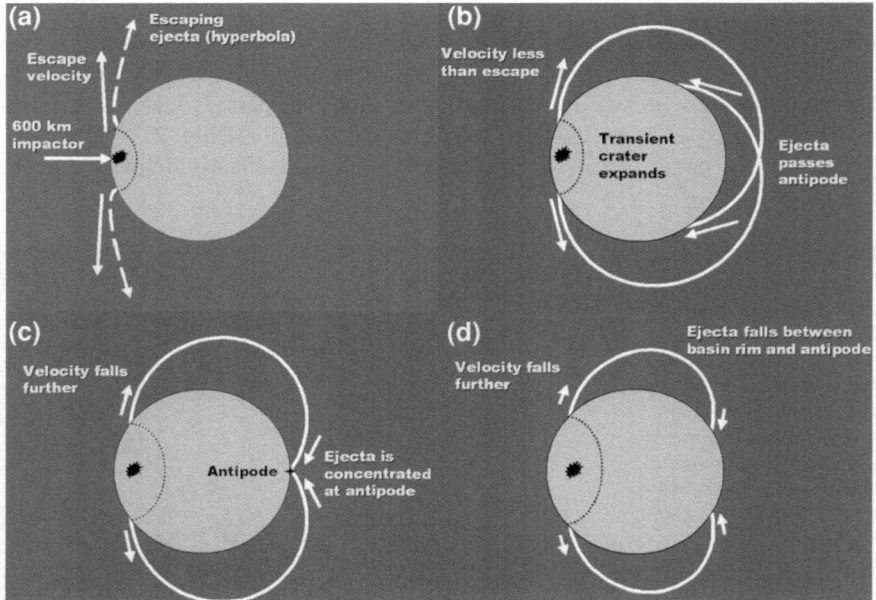

Fig. 6.14 This cartoon shows how two layers of NSM ejecta are laid down. Early ejecta escapes the Moon. As the radius of ejection expands the velocity drops below the escape velocity and the ejecta is thrown over the antipode but short of where the far rim will be, depositing the first layer. As the velocity drops further, the landing zone of the ejecta passes back to the antipode. It then recedes, laying down the second layer until it deposits the rim

Fig. 6.15 This shows a cross section of the far side bulge produced by the two layers of NSM ejecta, as they were deposited, before isostatic compensation

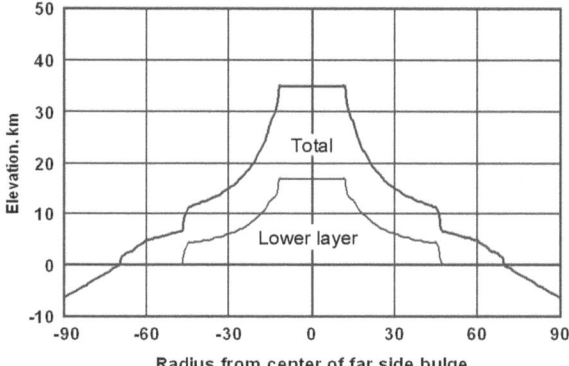

As the ejection proceeded, a second layer of ejecta was deposited over the first. The deposit zone receded until the rim was formed. There would be a scarp at the edge of the first layer as the trajectory of the ejecta passed from elliptical to hyperbolic.

Shortly after the two layers of NSM ejecta were deposited to form the far side bulge, before isostatic compensation, it was higher by a factor of 6 (Eq. 6.1) than it is now. Allowing for this ratio, the two layers of the far side bulge would be as shown in Fig. 6.15.

6.3.3 The NSM Subsurface Profile After Isostatic Compensation

Since there is a negligible long-wavelength free air gravity anomaly (Fig. 1.6) over the bulge (Neumann et al. 1996), essentially complete isostatic compensation has occurred. The surface and subsurface profile of the far side bulge as it is today is shown in Fig. 6.16.

The depth of each of the two layers is not reduced by isostatic compensation: they simply settle as units. Note that the thin incompatible layer between crust and mantle, the source of heat for eruption of molten mantle material, is also depressed. This probably accounts for the relative sparseness of maria on the far side bulge, especially outside of the SPA crater.

6.3.4 Future Analysis of the Far Side Bulge

The GRAIL mission, still in the lowest aktitude of its mission at this writing, may detect subtle differences in the two layers of NSM ejecta in the far side bulge.

Fig. 6.16 The surface and subsurface of the far side bulge as it is now, after full isostatic compensation. Recent analysis suggests a thinner crust (higher MOHO) than is shown here

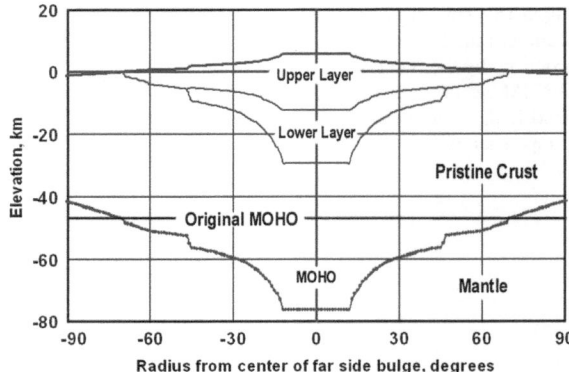

Such differences would be minimal near the NSM antipode where the bulge is greatest, but increase back toward the NSM rim.

If such a density variation were to be observed it would, however, be a counter-indication to recent observations of exposures of crystalline crust from various depths, which indicate a high degree of uniformity throughout the crust (Ohtake 2011). On the other hand, if no density variation is observed, it would but a constraint on the variation of crustal density with depth.

6.4   Age of the NSM

Identifying absolute ages of geologic features is not always easy on Earth but it is even more difficult on the Moon because of the limited samples and the difficulty of determining which feature is associated with a given sample. Because the NSM is such a gigantic feature and all of the lunar samples we have recovered were collected within it, there is a way of dating it. Recent work with zircon grains from two widely separated sites provide a pattern that supports a specific date for the NSM, just as similar patterned samples suggest a date of 3.9 Ga for the Imbrium Basin. The argument for an absolute age for the NSM follows.

6.4.1   Is the NSM the Oldest Megabasin? How Old Is It?

Events in the lunar pre-Nectarian period are of recent interest to several investigators (Losiak et al. 2009; Lineweaver and Norman 2009; Pidgeon et al. 2010). In this early period the solar system was settling down to its present state and life was starting on Earth (Lineweaver and Norman 2009).

Our best guide to absolute ages for early events is the study of lunar rock samples, based on concentrations of nuclei produced by radioactive decay. However, the strength of the bombardment has been such that most rocks near the surface have been through either a melt phase or impact shock, resetting their clocks. Fortunately, some of the samples contain zircon crystals, which retain their ages unless subjected to extraordinary conditions. All such rocks that have been returned from the Moon come from the flat floor of the NSM and provide evidence of the age of that event, particularly of the phase changes within its massive melt column. Until we have similar samples of the SPA and CM, both on the far side, we will not have absolute ages for those events.

Several investigators have concentrated on zircon crystals included in thin sections of breccia rock samples returned from the Moon by Apollo missions. These crystals are relatively refractory: they retain their clock settings at relatively high temperatures that melt other components of the same rocks and are also relatively resistant to shock. These crystals form from melts containing elevated amounts of incompatible elements, the source of KREEP. Although the principle compound that forms zircon crystals is zirconium silicate ($ZrSiO_4$), there is a significant component of uranium in the crystals, which slowly decays to lead.

A method of measuring the age of zircon crystals was suggested by Compston in 1977 and an age for an Apollo 17 sample (73217) was first reported in detail in 1984 (Compston et al. 1984; Heiken et al. 1991). The measurement is difficult because of the small size of the crystals (typically 10–30 μ) and the need to take measurements in even smaller spots in order to avoid the peripheral areas where some of the lead may have escaped. Consequently, it was difficult, in the early measurements, to obtain sufficient statistical evidence for precise aging.

Using a second-generation measuring instrument, (Kennedy and de Laeter 1994) to analyze zircons from several Apollo samples from two widely separated landing sites have provided precise ages for several events that have been strong enough to reset the zircon clocks (Nemchin et al. 2008, 2009). This leads to an interesting question: which events have been associated with which ages? The pattern of age determination from the two widely separated collection sites provides clues that associate certain events with specific age measurements.

6.4.2 Zircons in Thin Samples from Apollo 14 and 17

The rock samples analyzed in (Nemchin et al. 2008, 2009) were obtained from Apollo 14, in the ejecta blanket of the Imbrium Basin (Fra Mauro Formation) and from Apollo 17, in the Taurus-Littrow Valley just inside the southeast rim of the Serenitatis Basin. In both cases, the samples were thin sections of breccias, rocks that had been formed by impact shock applied to mixtures of smaller rocks from other sources. Measurements of the samples of both Apollo 14 and Apollo 17 showed discrete ages, with the ages of many individual zircon crystals correlated (Fig. 6.17).

The measurements of Fig. 6.17 were taken at three Apollo 14 stations and two Apollo 17 stations. Note that an age of 4.34 Ga was most frequent for each site, but there were several younger ages strongly represented at the Apollo 14 site. Yet all

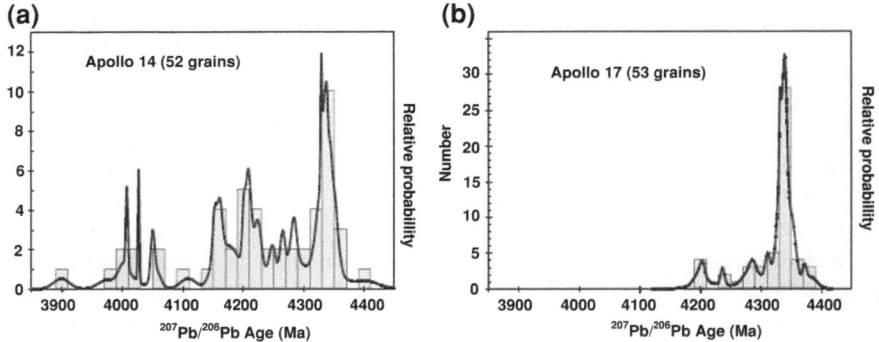

Fig. 6.17 Ages of individual measurements of zircons from samples taken from Apollo 14 and 17 sites (Nemchin et al. 2008, Fig. 10)

zircon ages except one were older than 3.9 Ga, the age of the Imbrium impact as determined from the ages of other minerals in the samples from this and other missions.

6.4.3 Where on the Moon did the Sampled Rocks Come From?

These rocks were taken from the surface of very different geologic contexts. The material at the surface of the Apollo 17 site near the rim of the Serenitatis Basin would have been brought up from a great depth. Impacts produce "overturn" whereby the surface of the rim is coated with materials that come from greater depths than the material from further out in the ejecta blanket. Surface samples could have come from the younger Imbrium Basin as well but they would not be as common as those from the depths below Serenitatis. If the material at depth was relatively homogeneous, that would explain the relative uniformity of the ages at the Apollo 17 site (Fig. 6.17).

On the other hand, the rocks taken from Fra Mauro Imbrium ejecta had been launched from well within the area that became the Imbrium Basin cavity. The Apollo 14 site (3.7°S, 17.5°W) is about 38° (great circle arc) from the center of the Imbrium Basin (34.0°N, 15.4°W). Figure 6.18 shows the relation between the internal radius (of launch) and the external radius (of deposit) for lunar ejecta. The apparent radius RI of the Imbrium Basin is 17.6°. Then the normalized external radius is 2.6 RI and the normalized internal radius, from the graph, is 0.74 RI.

Mare Imbrium now covers this position (about 22°N, 16°W) but of course was not there until it ejected material from the surface at that point to the Apollo 14 site. According to impact dynamics, that ejecta would have come from a much shallower depth than was the case with the Apollo 17 site and would also have included material from the target surface itself. Such material might not have been deeply buried in the chaotic Fra Mauro deposit event.

Fig. 6.18 The normalized radius of a deposit of ejecta is a function of the normalized radius of its launch. Normalization is on the apparent radius of the impact crater. The radius of deposit is the sum of the radius of launch plus the ballistic range implied by its velocity, at a typical launch angle of 45°. The launch velocity as a function of launch radius is derived in Chap. 4

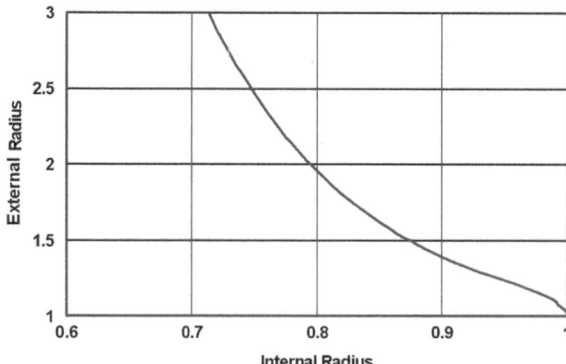

6.4.4 The NSM is 4.34 Ga Old

The zircons at age 4.34 Ga are too young to have retained the age of the primitive crust that solidified from the lunar magma ocean (Nemchin et al. 2008). These zircons come from rocks that had both deep and shallow sources and that were all collected within the flat floor of the NSM, the surface of its melt column. This melt column would reach far below the NSM cavity in a chaotic, turbulent process that mixed crust, mantle, and the incompatible layer. That process must have been the one that reset the ages of most of the zircon grains.

To summarize the evidence, the widely separated sites of Apollo 14 and Apollo 17 are each within the area of the level floor of the NSM and the dominant age of zircon crystals in both sites is 4.34 Ga. The deep source of the Serenitatis ejecta at Apollo 17 would have been from this re-melted crust and nearly all zircon ages are 4.34 Ga. The shallow source of the Imbrium ejecta at the Apollo 14 site must have brought up some of this re-melted crust with the most common 4.34 Ga zircons, mixed with material that had been re-melted (or very strongly shocked) by other impacts subsequent to the NSM but preceding Imbrium, unless they were deposited directly to the Apollo 14 site by post-Imbrium events.

6.4.5 Possible Impact Sources for Younger Zircons

The South Pole-Aitken Basin (SPA) was probably as effective as the NSM in bringing up zircons from the incompatible layer and resetting their clocks, but such zircons would have initially stayed within its cavity. Later impacts such as that of the Apollo Basin would have spread such zircons, but not too far from the SPA.

Zircons from the SPA and CM flat floors, collected near the rims of basins within them, would probably reveal their age, just as those collected from the flat floor of the NSM establishes its probable age. A caution here is that the SPA, NSM and CM impacts could have been a near-simultaneous barrage, so that the zircons from any two or all three of the events might have nearly the same age.

To explain the younger peaks of the Apollo 14 graph in Fig. 6.17, we should look close to the launch area within Imbrium of the Apollo 14 Fra Mauro deposit. For example, the Insularum Basin, revealed only by peaks of its rim rising above Mare Insularum, has its northern half actually buried by the rim and cavity of the Imbrium Basin. Zircons within the Insularum melt sheet may well have had their clocks reset. The northern edge of that melt sheet is very close to the launch point for the ejecta deposited at the Apollo 14 site; rocks from the northern edge of Insularum could easily have been thrown onto the Fra Mauro Apollo 14 landing site. The strong Apollo 14 peak at 4.16 Ga might be the age of the Insularum Basin. The 4.2 Ga age also appears at the Apollo 17 site and could be the age of a larger event than the Insularum impact. Several such sequences, starting with other basins, could have been the sources of zircon crystals with diverse ages, even within the same breccia sample.

Further analysis of zircon ages from Apollo samples for other landing sites would be very desirable. This could be done with existing thin samples or new thin samples from archived rocks. Newly recovered samples, especially from the two megabasins on the far side of the Moon would be especially valuable.

6.5 Summary of the NSM in Explanations of the Lunar Dichotomy

Properties of the NSM can be used to explain all of the evidences of dichotomy of the lunar surface and crust. A model of this single hypervelocity impact, quantitatively scaled from the models of the familiar basins and large craters, accounts for the large scale shape of the Moon. In the six years since the parameters of the NSM were published (Byrne 2007) data from new spacecraft and new techniques of simulation have confirmed and further explained the nature of the NSM. Recent evidence from the GRAIL mission has found that the highlands are at least 12% porous to a great depth (Wieczorek et al. 2012). This is in accordance with the far side bulge being formed primarily from ejecta from the NSM.

Also, patterns of minerals from rocky outcrops based on Kaguya spectorsopy have revealed concentrations in the overlap between the melt column of the NSM and the melt column of a possible Procellarum impact feature that impacted before the crust hardened (Nalamura, 2013).

Analyses of zircon grains in samples from additional Apolo landing sites have confirmed the pervasive age of 4.34 Ga for the NSM (Pidgeon, 2013).

Chapter 7
The South Pole-Aitken Basin

7.1 The Deepest Lunar Megabasin

The parameters of the South Pole-Aitkin Basin (SPA) are adjusted along with the NSM and other models and its characteristics are discussed. Identification of the SPA as a basin was late in coming (Wilhelms 1987). It was first suspected by examining Earth-based photography and finding mountains on the near side close to the South Pole (Hartmann and Kuiper 1962). After Zond, Apollo, and Lunar Orbiter provided clues seen by multiple observers, it was included in USGS geologic maps (Stuart-Alexander 1978; Wilhelms et al. 1979). The Clementine DEM shows most of the SPA well except for the southern part of its cavity.

Identification of the SPA as a basin was late in coming (Wilhelms 1987). It was first suspected by examining Earth-based photography and finding mountains on the near side close to the South Pole (Hartmann and Kuiper 1962). After Zond, Apollo, and Lunar Orbiter provided clues seen by multiple observers, it was included in USGS geologic maps (Stuart-Alexander 1978; Wilhelms et al. 1979). The Clementine DEM shows most of the SPA well except for the southern part of its cavity.

The SPA interacts with the NSM both by impacting the NSM ejecta at the far side bulge and by the conjunction of the rims of the two megabasins near the South Pole. Like the NSM, its rim is not clear because of the degree of erosion, shallow profile, and interaction with the far side bulge. Modeling the SPA and the NSM together clarifies the shapes of each.

This chapter concentrates on the SPA and its properties, examining the crater and its major features in detail. New understanding of the melt column, a result of new work with 3-D simulations on the SPA in particular is discussed. This work also sheds light on the SPA rim and ejecta field. Finally, there is a discussion of the age of the SPA in Sect. 7.6.

After the SPA was recognized, it was held to be the oldest and largest basin in the Solar System. It is neither. The NSM is older and larger, and the CM is older, as will be seen in Chap. 8 and Sect. 7.6 of this chapter. The Borealis Basin of Mars (Wilhelms and Squyers 1984; Andrews-Hanna et al. 2008) is larger than the NSM.

C. J. Byrne, *The Moon's Near Side Megabasin and Far Side Bulge*,
SpringerBriefs in Astronomy, DOI: 10.1007/978-1-4614-6949-0_7,
© Charles J. Byrne 2013

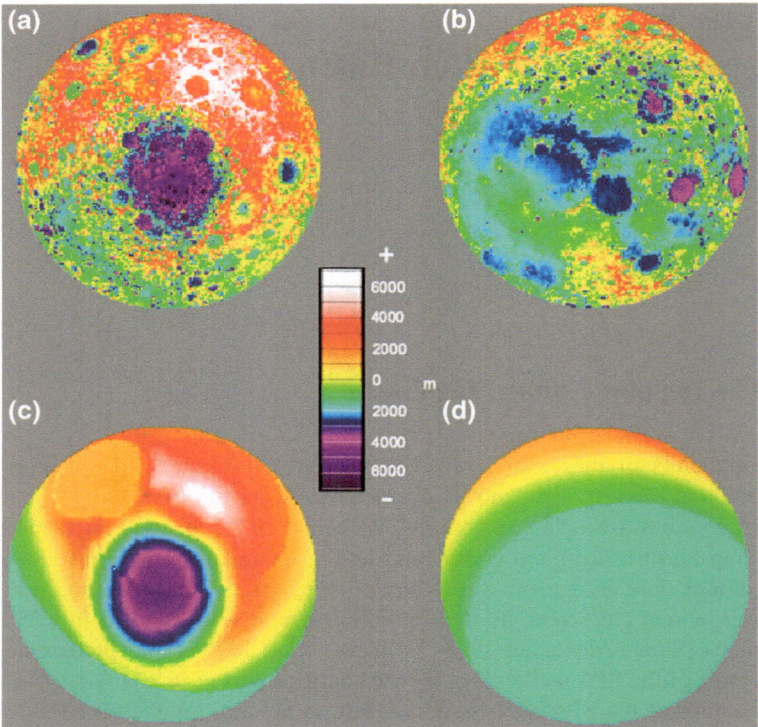

Fig. 7.1 Lambert equal area projections, 90° range, with centers on the SPA apparent cavity (*left*) and its antinode (*right*) . The Kaguya 1° DEM is shown in *a* and *b*. Corresponding views of the models of the NSM, SPA, and CM are shown in *c* and *d*. The sharp discontinuity in the SPA crater in the model is due to the scarp at the edge of the NSM lower ejecta layer and can be seen in the current topography

7.2 Focus on the SPA

The current topography of the SPA is compared to the analytic model of all three lunar megabasins in Fig. 7.1. Both current topography and the megabasin model are Lambert equal area projections (range 90°) centered on the SPA crater (48° S, 171° W) and its antinode (48° N, 9° E). The apparent crater of SPA is seen to be elliptical, indicating that it was produced by an oblique impact. This complicates the shape of its ejecta field as discussed in greater detail in Sect. 7.4. The apparent diameter is 3,000 km along the model's major axis and 2,400 km along the minor axis. The apparent depth, obviously the greatest on the Moon, is 7,050 m.

The center of the SPA model is 8° to the north of the common estimate in the past (Wilhelms 1987). The difference is probably due to the strong slope of the far side bulge from its high point north of the SPA apparent crater to the South Pole. The center of the SPA model is also 9° to the west of the common estimate, probably due

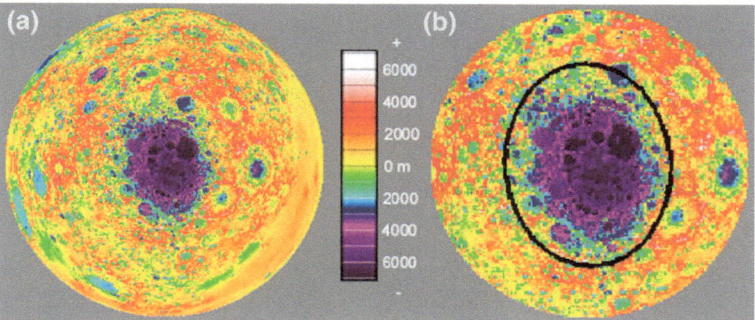

Fig. 7.2 a This map (Lambert 120°, centered on the SPA) shows the current topography with models of the NSM and CM subtracted to show the crater, rim, and ejecta field of SPA without distortion from the other megabasins. **b** This map (Lambert 80°) enlarges the view of the SPA crater and rim. The black ellipse is the modeled edge of the apparent crater, the intersection with the target surface

to the difficulty of determining the location of the much-eroded rims to the east and west. The model has a small SPA antinode deposit, but it is not observed in the current topography. It may have been absorbed in a still-molten NSM floor.

There is a strong discontinuity running east to west through the SPA model in Fig. 7.1c. This feature can also be seen in the current topography. It is due to the scarp of the lower layer of ejecta of the NSM at the radius where the ejected material is driven to escape the Moon. The remarkable effect on the SPA, confirmed by the current topography, was not seen until after the models' parameters were adjusted to minimize the standard deviation of the residual DEM. This feature is one of many details that confirm that the analytic model can be extrapolated to the size of the SPA and NSM.

In Fig. 7.2, models of the megabasins and the other large features discussed in Chap. 5 have been subtracted from the current topography, leaving SPA as by far the largest feature on the maps. This allows a view of the rim and ejecta field in Fig. 7.2a (Lambert 120° range) and a closer look at the crater in Fig. 7.2b (Lambert 80° range).

It's clear from Fig. 7.2a that both the rim and ejecta field are very uneven, probably distorted not only by subsequent bombardment but also by the oblique angle of the projectile. Figure 7.2b shows the elliptical edge of the apparent crater, which is also the inner edge of the rim. The floor of SPA, formed by the collapse of the melt column, is slightly south of the apparent crater's center; perhaps influenced by the slope of the target surface due to the pre-existing far side bulge.

The model of the SPA, shown with the other two megabasins in Fig. 7.1c, was modified to accommodate the observed elliptical shape. The method was to allow the apparent radius to vary according to the eccentricity and angle of the major axis. The same method was used with the NSM.

Figure 7.3 shows the comprehensive residual DEM with the same projection as Fig. 7.2a.

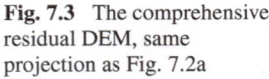
Fig. 7.3 The comprehensive residual DEM, same projection as Fig. 7.2a

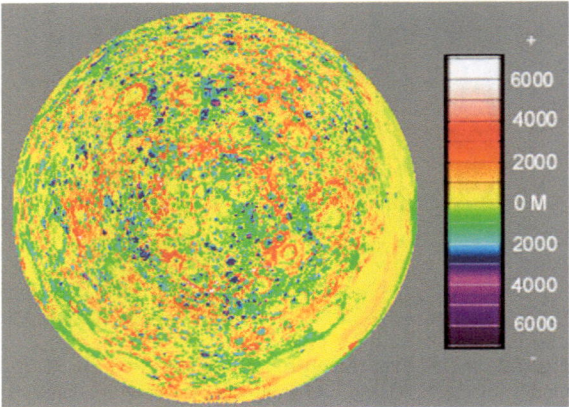

The interior of the SPA apparent crater is well modeled, showing very little error in Fig. 7.3 but there is some residual deviation at the rim, as might be expected for an oblique impact. There is also a suggestion of arcs of an external ring (this ring was not modeled).

7.3 The Apparent Crater of the SPA

Although we do not have returned samples from the SPA apparent crater and no meteoroids associated with the Moon appear to be correlated with remote sensing data of that area, recent advances in remote sensing and simulations have provided a new understanding of its apparent crater.

The following subsections describe the range and depth of the NSM crater, the effects of isostatic compensation and crustal thickness, and the floor and the melt column of the SPA crater.

7.3.1 Range of the Apparent Crater of the SPA

The model of the apparent crater of the SPA extends from its center at 48° S and 171° W to 4° N latitude on the far side southward beyond the equator and the South Pole to 8° S on the near side. In longitude, it extends from 124° E to 106° W. Its major axis is tilted 4° counter-clockwise. The SPA center is northwest of the center given in "The Geologic History of the Moon" (Wilhelms 1987) and the apparent crater is considerably larger than that estimate. Early estimates of these parameters were derived from oblique photographs by Lunar Orbiter; the differences may be due to the low slopes of the deep but shallow crater, erosion by bombardment, and the unavailability of comprehensive elevation data.

The model reported here is based on complete DEM coverage of the SPA, removal of a model of the NSM that produced the far side bulge, and averaging over azimuth that smoothes interference from bombardment. This model not only is a formal best fit to the topography but also, when removed from the current DEM along with other models, makes the SPA nearly disappear from the residual map of Fig. 7.3.

7.3.2   Depth of the Apparent Crater of the SPA

Orbital traverses by Apollo missions equipped with radar altimetry established that there was a deep cavity within the far side bulge. Only after mountain ranges (arcs of the rim), were identified was it possible to associate the cavity with a probable impact crater (Stuart-Alexander 1978; Wilhelms et al. 1979; Wilhelms 1987). Even the Clementine DEM was not helpful south of latitude 79° S because of the orbital altitude there being beyond the range of the laser altimeter. Currently, we have precise elevation maps from recent spacecraft missions that cover the entire SPA crater, rim, and ejecta field, allowing us to generate a radial profile of the entire feature (see Fig. 7.4).

Qualitatively, the agreement between a circular SPA model (eccentricity set to zero and diameter set to 2,684 km) and a radial elevation profile (with the other megabasins removed) is reasonable but with some variations. The rim appears to have been badly eroded, and perhaps eroded material has fallen into the crater. The model shows a deposit at the antinode that does not appear in the current topography. The missing antinode could be due to the oblique impact or could be that the floor of the NSM was still molten crust when that deposit arrived. The raised

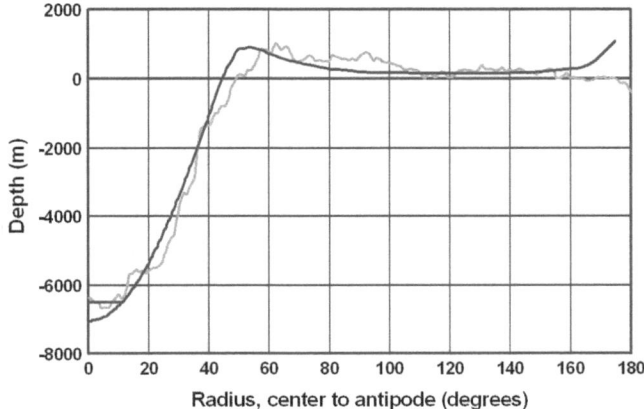

Fig. 7.4 Comparison of a circular model of the SPA (*black line*) to the radial average of the current elevation (*gray line*) from the center of the SPA to its antipode. All other models have been subtracted from the current topography to isolate the radial profile of the SPA

elevation at a radius of about 90° is probably due to an external ring of the SPA, also visible in the residual DEM of Fig. 7.3.

The scaling depth of the current, post-compensation crater is 7,050 km. That the model level of the floor (set by parameter variation to minimize standard deviation) is slightly lower than the observed floor is probably due to the actual floor being to the south of the center of the apparent crater. This is also the probable cause of the discrepancy between earlier estimates of the center latitude of the SPA and the model estimate: the earlier estimates may have been set at the center of the flat floor, while the model estimate compromises between the flat floor and the rim of the apparent crater in order to optimize the agreement between the model and the current topography.

7.3.3 Isostatic Compensation, Gravity, Crustal Thickness, and Ejecta Volume

The SPA, like the NSM, has undergone isostatic compensation. The interaction between the two megabasins is discussed fully in Sect. 6.2.3. There are negligible free-air gravity anomalies that reflect the topography of either the crater of the NSM or its ejecta field in the far side bulge, establishing that it has undergone complete isostatic equilibrium. However, there is a small negative free-air gravity anomaly over the SPA crater (see Fig. 7.6), indicating that its isostatic compensation may be incomplete, considering the presence of some mare.

The apparent depth of the SPA model (after isotopic compensation) is 7,050 m: but the initial apparent depth would have been 42.3 km if the SPA is essentially fully compensated. That depth would have been established for only a short interval, because the SPA melt column (which would have penetrated to a much greater depth, far into the mantle) would have expanded and then collapsed to the floor. The heat imparted by the collapsed melt column would have expedited the compensation process, raising parts of the mantle, even as the melt column collapsed (see the simulation in Fig. 7.10). The deepest part of the modeled floor is now at an absolute depth of 5,700 m below the reference geode.

The total volume of material ejected from the original apparent crater would have been 116 million cubic kilometers (see Fig. 7.5). As the ejection cone expanded to a normalized radius of 0.66, 81 million cubic km of ejecta escaped from the Moon. As the ejecta cone approached the rim, an additional 35 million cubic kilometers of ejecta would have been deposited on the rim and ejecta field.

The maps in Fig. 7.6 through 7.9, from Kaguya data. (Sasaki et al. 2011) that presents a coherent set of figures that have been published in various forms in several other publications. A map of free-air gravity is shown in Fig. 7.6.

A free-air gravity anomaly may be due to uncompensated topography or to a change in density. Either an uncompensated crater in a target of uniform density or an area of low density in flat topography decreases the local gravity field. In the case of the SPA, a lower density over its melt column seems unlikely, so partial compensation of the crater is more reasonable.

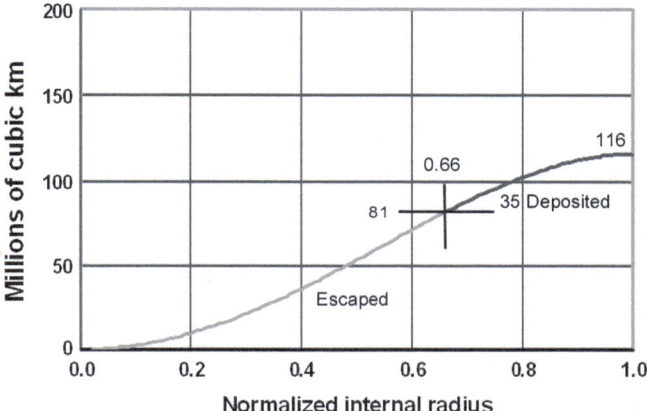

Fig. 7.5 Cumulative volume of ejecta from a vertical impact model (*circular*) of the SPA as a function of the normalized internal radius of its apparent cavity. Of a total 116 million km ejected from the apparent cavity, 81 million cubic km escaped the Moon and 35 million cubic km was deposited in the ejecta field

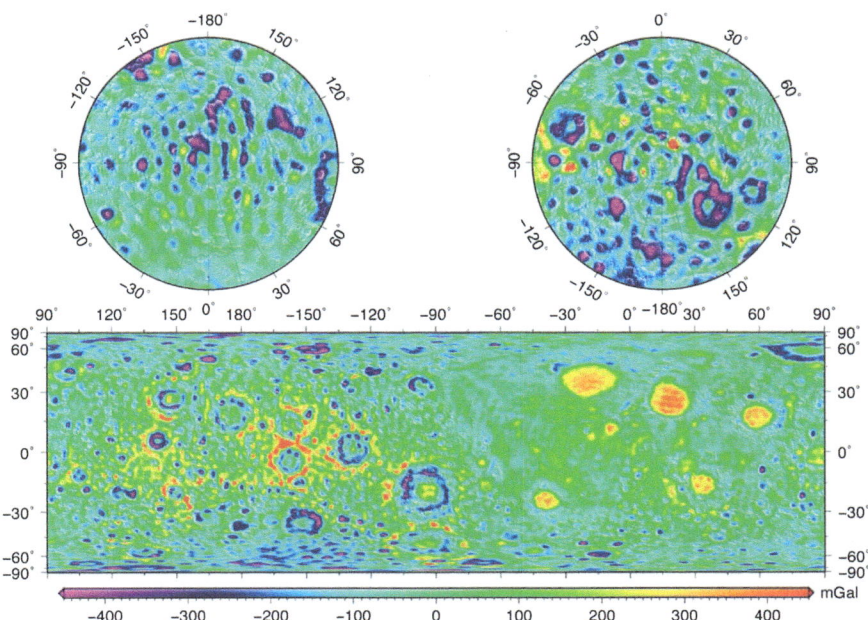

Fig. 7.6 Free-air gravity map of the Moon. The latitude scale is adjusted to show equal areas for the features, although their shape is distorted in the lower map. The north polar region is shown on the *upper left* and the south polar region on the *right*. While there is no significant anomaly over the crater of the NSM, there is a small negative anomaly over the SPA (Sasaki et al. 2011)

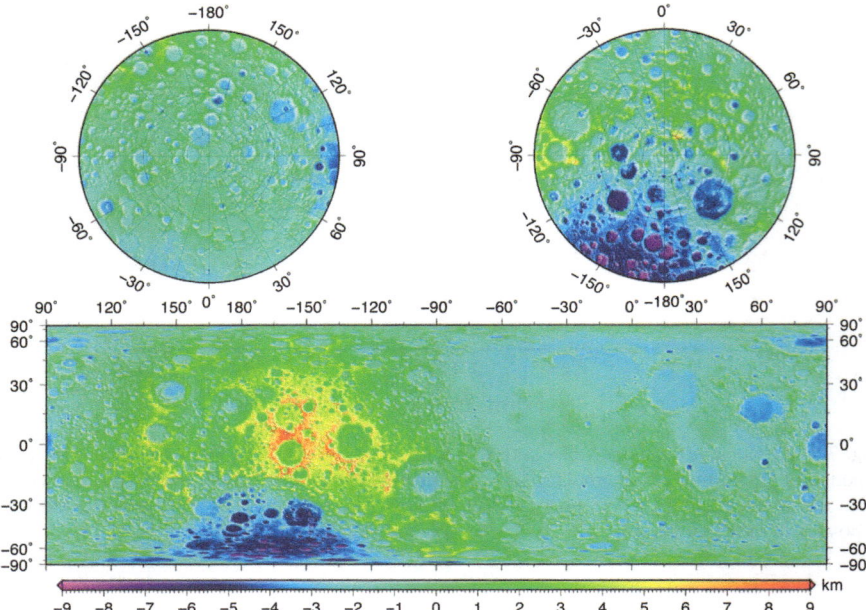

Fig. 7.7 Topography map of the Moon in the same projection and scale as Fig. 7.6. This is a representation of the same Kaguya data as has been used in previous chapters for current topography except for differences in the latitude scale, the false color scale, and higher resolution (Sasaki et al. 2011)

Bouguer gravity combines topography with free-air gravity. Variations in gravity due to the measured topography are removed. Therefore the residual anomalies reflect variations in density and the depth of variations in density. Types of such variations are:

- Porosity of ejecta relative to the target surface
- Mare deposits of basaltic lava
- Uplift of basaltic mantle to compensate for variations in topography.

Figure 7.8 shows a positive Bouguer anomaly in the SPA, analogous to the mascons in the NSM crater. This could be due to mare basalt at or just below the surface (cryptomare). Note the iron anomaly in in the SPA shown in Fig. 7.12. The positive anomaly of the far side bulge in Fig. 7.7 is more likely due to high porosity (low density) of the ejecta from the NSM.

In Fig. 7.9, the crustal thicknessof the SPA is shown to be thinnest in the region of the melt column but grades upward toward the rim. Despite the great depth of the SPA crater, the crust is thinner under the Crisium Basin and the small Moscoviense Basin. New analyses from GRAIL data (Wieczorek 2012) indicate a thinner crust than shown here.

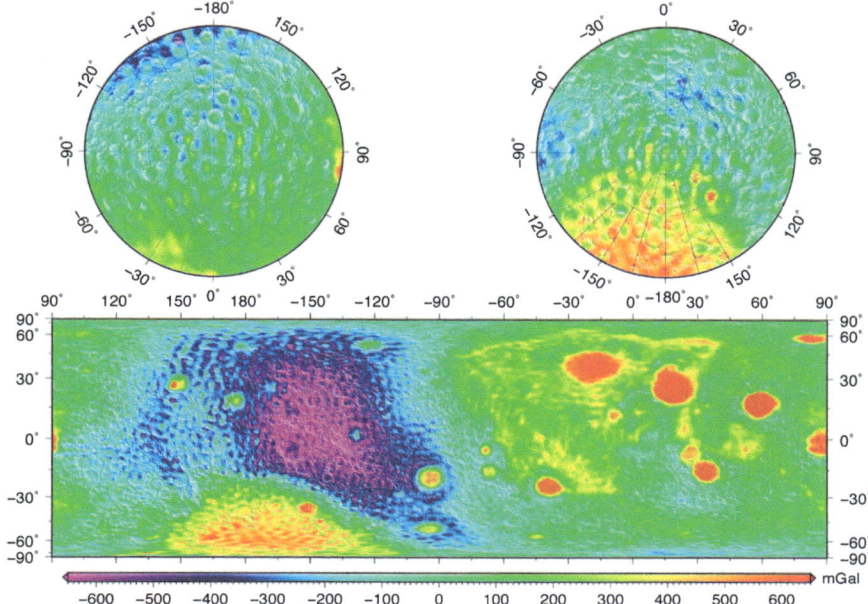

Fig. 7.8 Bouguer gravity map produced by combining the data of Figs. 7.6 and 7.7. There is a strong positive anomaly over the SPA, especially over the region of its melt column. There is an even larger negative anomaly at the far side bulge (Sasaki et al. 2011)

7.3.4 Description of the Floor and Melt Column of the SPA

Recent three-dimensional simulations of the formation of the SPA (Ivanov 2007; Stewart 2011, 2012a, b) with both normal and oblique impacts provide a deep insight into not only this great impact but also, through its example, the formation dynamics of all such basins. In the megabasins, most of the shock wave energy is converted to phase changes in the target material far beyond the crust into the mantle. The surface manifestation of a megabasin, the apparent crater, rim, and ejecta blanket, is the visible evidence of the feature, whose influence on the target body extends far below the surface and crust. Nevertheless, the simulated dynamics of the surface features, except for the flat floor, have the same observed self-similar shapes as any other impact crater.

Single-frame excerpts from the movie of one of S. T. Stewart's simulation are presented in Fig. 7.10.

In the simulation of Fig. 7.10, the projectile and crust are partly pulverized, partly molten, and partly vaporized. By 500 s, some of the debris explodes upward and downstream as parts of the projectile and crust penetrate the crust and are driven deep into the mantle. By 2,000 s, some of the crust is thrown outward and upward toward an ejection field while mantle material rebounds upward and mixes with crust and projectile material. By 4,000 s, the crust and projectile material that was ejected vertically collapse into the crater. Since the material is mostly molten,

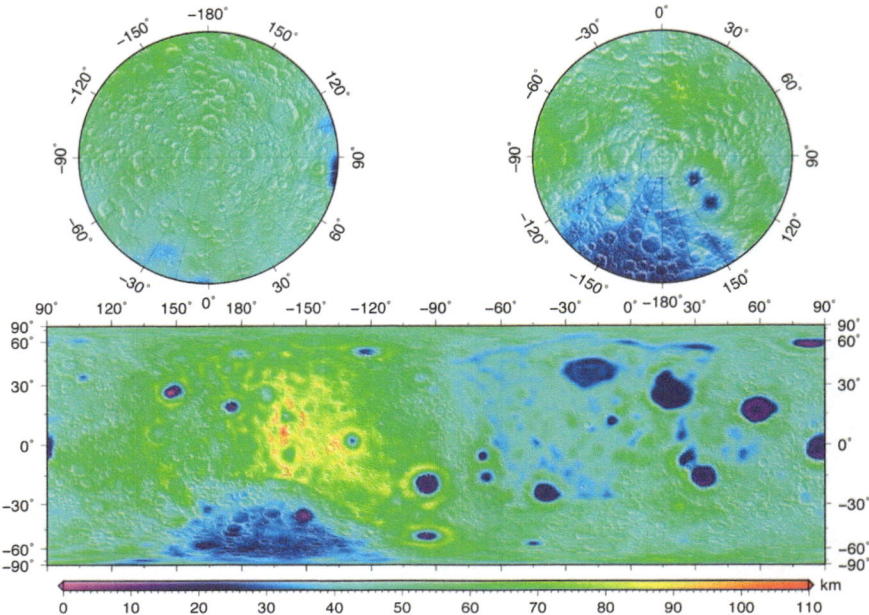

Fig. 7.9 Crustal thickness inferred from the Bouguer anomaly and assumptions of density of the crust and mantle (Sasaki et al. 2011)

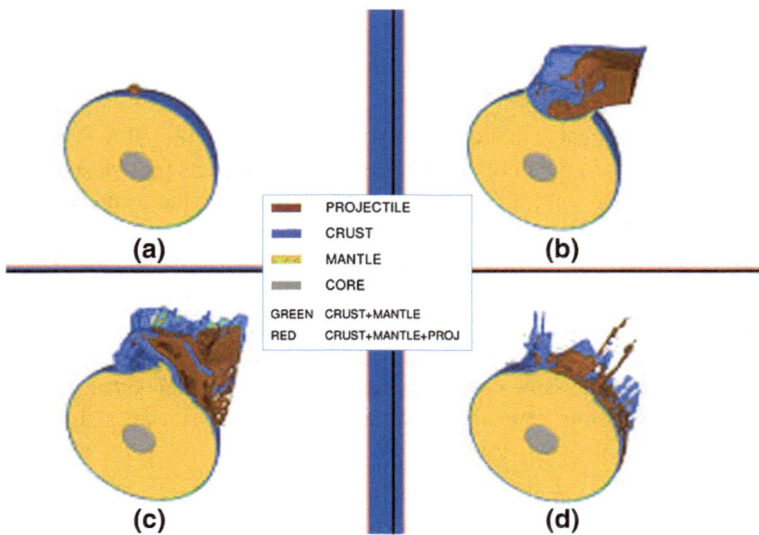

Fig. 7.10 These frames are from a movie of a 3-D simulation of and SPA impact, with the projectile approaching at an angle of 45°. **a** Impact, 0 s. **b**. Ejection of crust and penetration into the mantle, 500 s. **c**. Mantle rebounds and ejection continues, 2,000 s, **d** Mantle settles into flat floor, covered by crust. Some projectile material covered by crust settles downstream from impact to rim and near ejecta field. 4,000 s. *Source* (Stewart 2012b)

it would form a flat floor of the crater. In a vertical impact on a target surface that follows the geode, the material from the melt column forms a flat floor roughly the diameter of the melt column, but in the case of the SPA's oblique impact on a sloped target surface, the floor shape is more complex. The ejected material is deeper on the downstream side of the oblique impact.

7.4   Rim, Ejecta Field, and Direction of Impact of the SPA

As with the melt column, the simulations provide insight into the formation of the rim and ejecta field of oblique impacts and the SPA in particular. A plan view of a simulation of the SPA (Stewart 2012a) after the dynamics subside is shown in Fig. 7.11.

The temperature map of Fig. 7.11 shows that the ejected material is cool; the shock energy near the surface is nearly all converted to kinetic energy. Therefore, the extended Mazwell-Z model developed in Chap. 3 applies, except for the

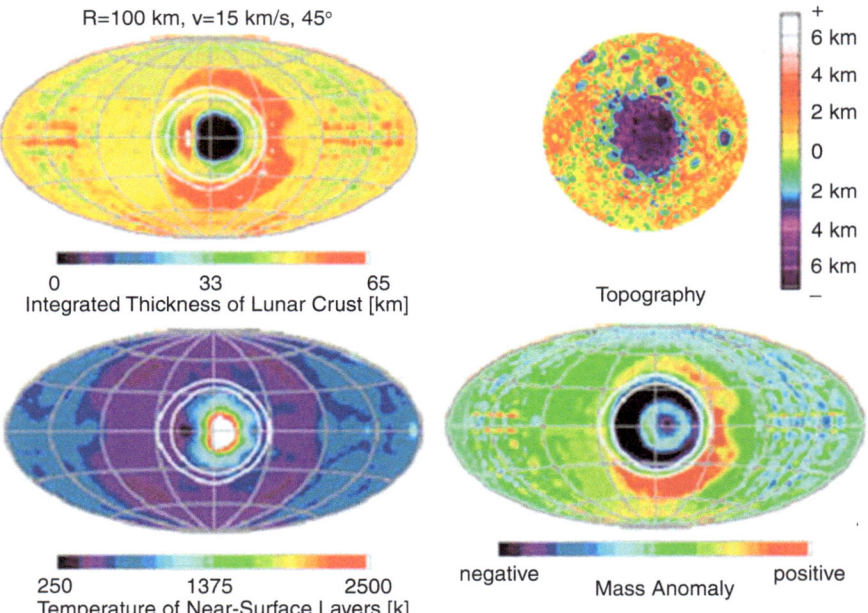

Fig. 7.11 Horizontal view of the result of the simulation of the SPA (Stewart 2012a) under the conditions of Fig. 7.10. The current topography of the SPA is on the upper right, Lambert 90° to match the scale of the other maps current author) as in Fig. 7.2. The maps on the left and bottom right (Stewart, 2011) show a horizontal view of the simulation of Fig. 7.10. The map on the top right (current author) shows, for comparison, the current topography of the SPA with models of other features removed as in Fig. 7.2 (Lambert 90°, same scale as other maps of this figure). Note that the rim and ejecta field are more extensive to the East coincidentally aligned with the downrange direction of the simulation. The scales of the thickness and mass diagrams apply to both time of impact and current conditions but the current topography is much less than at the time of impact because of isotopic compensation

effects of the oblique impact. The intensely heated material is mostly confined to the vertical melt column which explodes upward and mostly collapses on itself.

The topography map of Fig. 7.2 shows the current DEM of the Moon, with models of the NSM, the CM, and the mounds and depressions described in Chap. 6 removed to show the SPA without interference. Fortuitously, the major and minor axes of the SPA align approximately with the simulation. The floor of the SPA (the blue and purple colors) reflects roughly the shape of the simulated impact. Apparently due to the oblique impact of the slope of the far side bulge, the floor of the SPA in the area of the melt column is not quite flat as it is in the NSM. The SPA ejecta layer is deeper to the right, the downstream direction of the simulation where the post-impact crust is thicker.

Overall, the shape of the apparent crater is elliptical, with the major axis at right angles to the direction of impact, as predicted by terrestrial experiments conducted by Peter Schultz (Schultz 2007). Comparison of the simulation with the topography suggests that the direction of impact of the SPA was approximately West to East, somewhat of a surprise.

While the crustal floors of megabasins have been seen in the past (Byrne 2007), the new paradigm of the melt column formation and evolution explains how they form, while preserving the previous model of the slope from the floor to the rim and the ejecta field. It has been suggested that similar but lesser phenomena may occur in the larger basins such as Orientale, as the melt pool of the smaller basins evolves toward the melt column. This would be an interesting topic for more simulations. Simulations of the NSM and CM megabasins are also eagerly anticipated.

7.5 Mineral Anomalies of the SPA

The floor of the SPA, like the floor of the NSM, is rich in minerals that have more heavy metals than the anorthositic crust. That comes, of course from the chaotic mixing of crust and mantle and the thinning of the crust. The range of

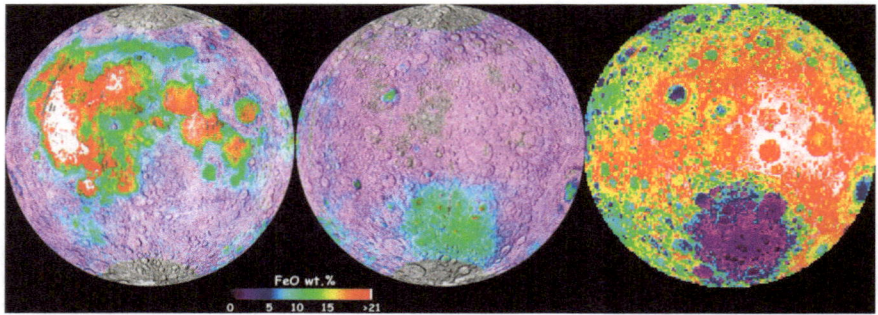

Fig. 7.12 *Left* Iron concentrations on the near side of the Moon. The high iron concentrations are on the flat floor of the NSM. *Center* Iron concentrations on the far side of the Moon. The iron concentrations are nearly limited to the floor of the SPA. *Source* NRL, Clementine, Jeff Gillis: http://meteorites.wustl.edu/lunar/moon_meteorites.htm. *Right* Topography of the far side of the Moon. False color map for elevation as previous maps

the iron content of the surface of the floor of the SPA is compared to the topography in Fig. 7.12.

Comparison of the center and right images in Fig. 7.12 establishes the area of the melt column where mantle material has mixed with crust chaotically before settling and re-crystalizing. The SPA is much smaller than NSM, so the mechanism of enhancing the heavy metal content is somewhat different. The crust was entirely reformed in the NSM and then largely covered with basaltic lava from the mantle perhaps 500 million years later. But the SPA melt column, being smaller, must have re-crystalized much quicker and may well have retained mantle inclusions near the surface that were excavated by subsequent heavy bombardment. Some lava was erupted in the SPA much later, just as with the NSM but not nearly as much.

7.6   Age of the SPA

The SPA is widely believed to be younger than the far side bulge and therefore than the bulge's cause, the NSM. Here, the shape of the SPA has been derived from superposition which shows a strong interplay between the two layers of the NSM ejecta and the shape of the SPA. Unfortunately, that does not resolve the sequence. The strong mineral anomalies of the SPA melt column do provide evidence that the NSM was first: otherwise, much of the anomalistic material would have been obscured by the NSM ejecta, although some would have been excavated by bombardment. So the SPA is probably younger than the NSM (SPA), which was estimated in Chap. 6 to be 4.34 Ga old.

How much younger? That's hard to say. The SPA has been impacted by as many large pre-Nectarian craters as any other pre-Nectarian surface. Therefore, as previously estimated (Wilhelms 1987), SPA happened very early in the pre-Nectarian period. Unfortunately, a precise absolute age for the Nectaris Basin has not been established. So we can only say, in absolute terms, that the SPA is younger than 4.34 Ga and much older than the Imbrium Basin, whose age is about 3.9 Ga. The extensive pre-Nectarian bombardment impacts in the SPA is evidence that its age cannot be very close to that of Nectaris, but there is no evidence that its age could not be close to that of the NSM.

It is possible that the missing antinode deposit of the SPA may have been absorbed in the still-plastic melt column of the NSM (See Fig. 7.4 and the accompanying discussion). If that is true, then the SPA must have been formed before the NSM melt column would have hardened, a time that must have been less than but comparable to the time that was required for the Lunar Magma Ocean (LMO) to harden, estimated to be of the order of 100 million years (Shearer et al. 2006), although that age is controversial. The age of the SPA would likely have been less than the age of the NSM by not more than about 100 million years. Then the age of the SPA would be between 4.34 Ga and 4.24 Ga.

Chapter 8
The Chaplygin-Mandel'shtam Basin

8.1 The Smallest Early Lunar Megabasin

The Chaplygin-Mandel'shtam Basin is a newly identified megabasin that is a smaller megabasin that is intermediate in age between the NSM and the SPA: it is revealed by subtracting other models from the current topography. The NSM and SPA were modeled together with the models of the other basins and large craters. The intermediate residual DEM was examined for remaining features. One of those was a depression in the northeast region of the far side, larger than the Imbrium Basin but smaller than the SPA. Although hard to identify in photographs and subtle in the current topography, a classic model of an impact feature was revealed in the radial profile. The named craters Chaplygin and Mandel'shtam are within its apparent crater.

The shape of the Chaplygin-Mandel'shtam Basin (CM) (floor, rim, and ejecta field) is visible at the surface (highly eroded), but that would result from superposition of later ejecta. Figure 8.1 shows an interim residual DEM that reveals the CM, the model of the CM, and the residual DEM after including that model.

The CM interacts with the NSM rim and ejecta field on the slope of the far side bulge. After the CM was identified by the residual in the current topography its parameters were estimated by the radial profile method. The NSM parameters were then recomputed in the presence of the CM model. Finally, the CM parameters were readjusted to reduce the standard deviation of the final residual DEM. Because the CM is much smaller than the NSM, the interaction is restricted to a relatively small area but it is strongest where the CM and NSM rims become close west of the Moscoviense Basin and the Mendeleev Basin.

There are two ways the CM and the NSM can interact (see Fig. 8.2), depending on whether the CM or the NSM impact happened first.

If the CM is the first and oldest of the megabasins, it would have encountered a nearly pristine crust, a gravitationally level surface. Consequently, the collapsed melt column would have achieved a flat, level surface. The NSM ejecta would have fallen into the CM cavity in that area, resulting in the profile shown in Fig. 8.2.

C. J. Byrne, *The Moon's Near Side Megabasin and Far Side Bulge*, SpringerBriefs in Astronomy, DOI: 10.1007/978-1-4614-6949-0_8,

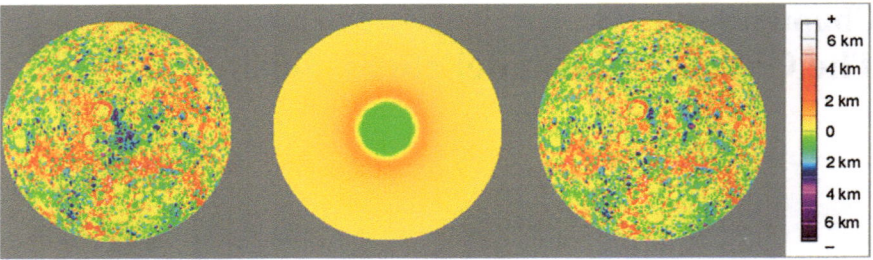

Fig. 8.1 *Left* The interim residual topography (Lambert 90°, centered on CM) with the other large features removed but without the CM. *Center* The model of the CM (the model of the ejecta layer of the CM is positive beyond the rim but is obscured by the color scale). *Right* the residual DEM after including the model of the CM

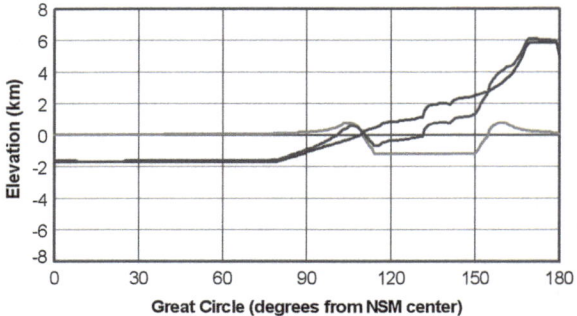

Fig. 8.2 The interaction between the current topography of the CM and the NSM is shown along a great *circle* that connects the centers of the two megabasins with the angle proceeding toward Moscoviense. The model of the CM is *red*, the NSM model is in *blue*, and the superposition of the two is in *black*. There are two ways to interpret this model, depending on whether the CM impact preceded or succeeded the NSM event

On the other hand, if the NSM is the earlier megabasin, its ejecta would have encountered relatively even crust in this area (at least at a large scale), just as it has in the rest of the far side bulge. The CM would have impacted the sloped surface of the far side bulge just as has the SPA. It has been suggested (Wilhelms 2012) that the CM is only a little larger than the Imbrium Basin. Therefore why distinguish the CM as a megabasin? The point is significant. By the rule used for Mars (see Chap. 2), the rule distinguishing megabasins is an apparent diameter of 1,000 km. The Imbrium basin would pass that test. The CM includes younger basins (Freundlich-Sharanov for example), while any other possible basins within the Imbrium Basin are obscured by Mare Imbrium. Could the Imbrium Basin have had a melt column? Again, a melt column of the Imbrium Basin would be obscured by Mare Imbrium. In the case of the even smaller Orientalis basin, a melt column has been suggested (Stewart 2011; Andrews-Hanna 2011; Vaughan 2012), one that rose well above the current floor within the inner Montes Rook ring, even splashing mantle material onto the terraces, and then subsiding to the level of Mare Orientalis.

In any case, there is another distinction between the Imbrium Basin and the CM, that of age. The CM is not only older than the Moscoviense, Mendeleev, and Korolev Nectarian Basins but also older than the pre-Nectarian Freundlich-Sharanov Basin. The Imbrium Basin is much younger than all of these. Age alone does not disqualify the Imbrium basin from being a megabasin, even though other distinguishing properties are not visible, but age does suggest assigning the NSM, SPA, and CM in a separate age group from the large Nectarian and Lower Imbrium basins. Pehaps they should be called early megabasins.

In Chap. 10 on the history of the Moon, the age distinction is discussed further, attributing the Early Heavy Bombardment (EHB) to planetesimals in the early solar system and the Late Heavy Bombardment (which could be found to have produced megabasins) to a disruption of the residual smaller planetesimals in the asteroid belt. Such a late disruption is inferred from the Nice Model.

The relative ages of the CM, the NSM, and the SPA cannot be positively distinguished by superposition of the elevation models alone. Residual standard deviations were determined for these cases: CM first, CM between the NSM and the SPA, and CM after the SPA. The best fit to the current topography was found to be the case where the CM followed the NSM and preceded the SPA. In this case, The CM levels its flat floor and then ejecta from the SPA falls on the CM from the south, restoring some of the slope of the far side bulge there.

The questions of whether the Imbrium Basin and others of the late heavy bombardment had a melt column and internal basins are open. Perhaps GRAIL will provide additional evidence of the subsurface structure to support answers to these questions.

8.2 Focus on the CM

The place of the CM in the current topography is shown in Fig. 8.3. Note the arc of high elevation to the southwest. The surface there is probably not the directly exposed rim of the CM but overburden from later features.

8.3 The Apparent Crater of the CM

The apparent crater is buried below ejecta from many later basins. That material has been further disturbed by the subsequent history of bombardment. Consequently, the apparent crater is not so much viewed as inferred from its effects on the layers superposed on it.

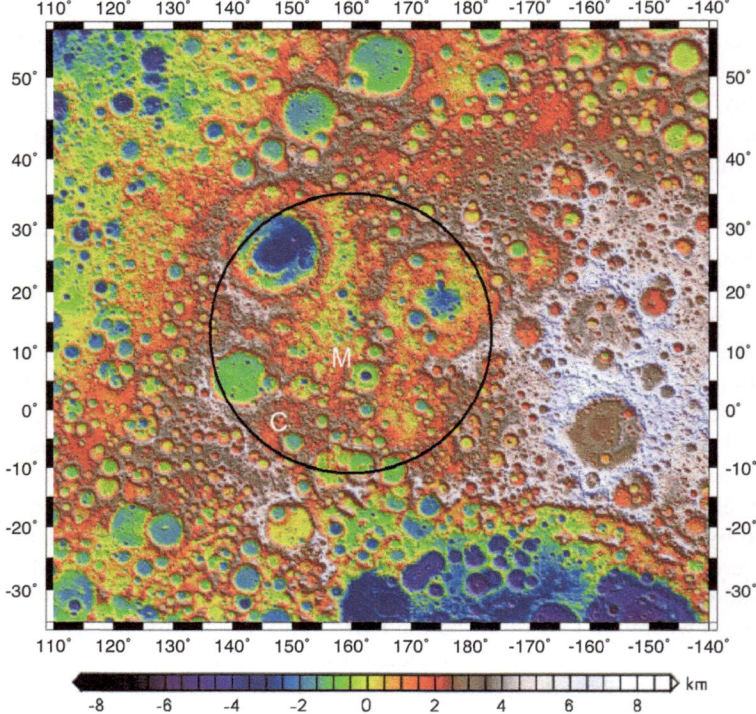

Fig. 8.3 Current topography of the area of the CM megabasin. The *black circle outlines* the apparent crater of the CM. Chaplygin is marked C and Mandel'shtam is marked M. The CM crater, rim, and ejecta field extends across nearly all the slope of the far side bulge. Note the elevated *arc* to the southwest and south, evidence of the rim in the shallowest part of the NSM ejecta field. m. *Source* NASA, LRO, Goddard Space Flight Center, (Neumann 2012)

8.3.1 Range of the Apparent Crater of the CM

The CM is centered at 9.6° N and 162° E and has an apparent diameter of 1,320 km (see the elevation and shaded relief map of Fig. 8.3). Its rim encloses parts or all of the Moscoviense, Mendeleev, and Freundlich-Sharanov basins, as well as the recently identified Kohlschutter-Leonov Basin (Cook 2002; Scholten 2011). Parameters of these basins are shown in Chap. 4, Table 4.2. The western part of its rim intersects the eastern rim of the NSM. Another view of the area, with the CM megabasin and neighboring basins outlined, is shown in the photograph of Fig. 8.4.

The current surface above the apparent crater, rim and ejecta blanket of the CM is an interesting study in stratigraphy. Basins from the pre-Nectarian and Nectarian periods are superposed on the CM and SPA megabasins. The degree of cratering

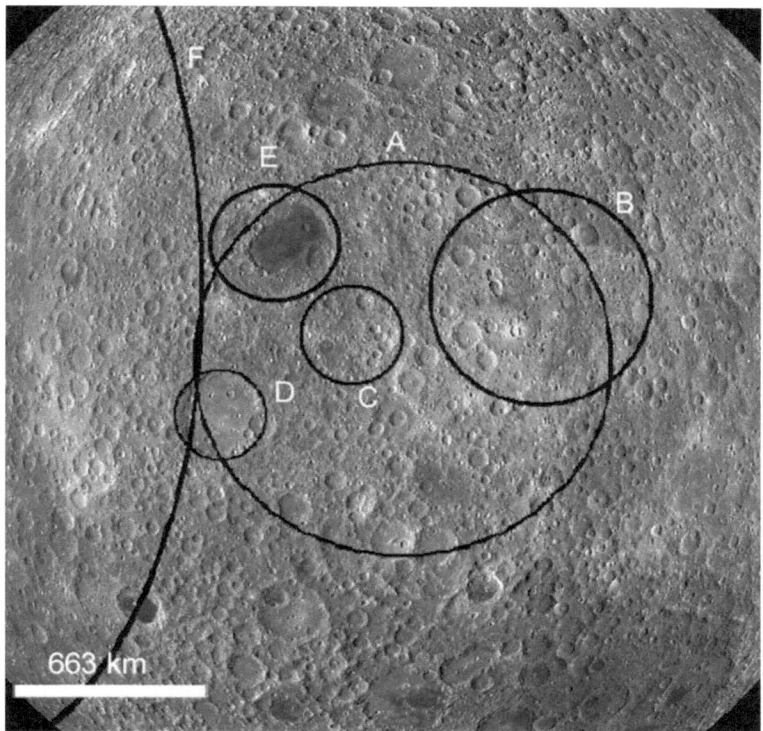

Fig. 8.4 This photographic mosaic from the LRO Wide Area Camera shows the interactions of the following impact features; (*A*) CM, (*B*) Freundlich-Sharanov), (*C*) Kohlschutter-Leonov, (*D*) Mendeleev, (*E*) Moscoviense, and (*F*) NSM. The *black circles outline* the apparent craters. The *dark area* to the southeast is the SPA. Source of the base image: NASA, LRO, Arizona State University (LROC 2012)

in different regions, with older craters being obscured by ejecta blankets from younger neighboring basins, reveals the sequence of the features. In particular, note the difference between the high density of superposed craters on the apparent crater, rim, and ejecta field of the Freundlich-Sharanov Basin (pre-Nectarian age group 8) and the relatively low density in the similar structures of the Moscoviense Basin (Nectarian, age group 1 and the younger Mandeleev Basin (Nectarian age group 2). These differences reveal the degree of bombardment in the intervals between those impacts.

In the western CM apparent crater, the relatively recent and overlapping ejecta from the Mendeleev and Moscoviense basins cover the older surface, obscuring smaller pre-Nectarian craters, as noted in geologic maps (Stuart-Alexander 1978). This can be taken as evidence of the amount of impacts that occurred between the impact of the pre-Nectarian Freundlich-Sharanov Basin and the later Nectarian basins.

8.3.2 Depth of the Apparent Crater of the CM

Figure 8.5 compares the model of the CM megabasin to the surface topography (Kaguya 1/16° DEM) with the models of the other megabasins and large features removed. Some of the shape of the rim of the CM can be seen in the false color image in Fig. 8.3 (the pink arc to the southwest), even though it is covered with overburden. The apparent depth is 3,800 m and the fill is 1,200 m below the target surface.

8.3.3 Isostatic Compensation and Crustal Thickness of the CM

Like the NSM, the CM has no free-air gravity anomaly (see Fig. 7.6). Consequently, it must have undergone full isostatic compensation, as well as the NSM ejecta below it. Figure 8.6 shows the post-compensation cross section of the NSM, the CM, and the lunar Moho along the great circle that connects their centers. A depth of the initial Moho of 47.5 km (Byrne 2007) was assumed in Fig. 8.6.

Fig. 8.5 The surface expression of the CM megabasin (with models of the NSM and SPA and other large features removed) compared to the CM model

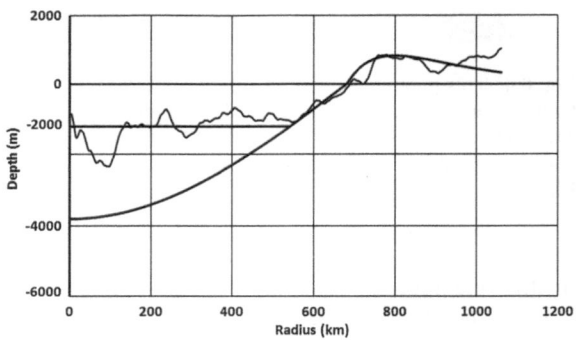

Fig. 8.6 Cross section of the NSM, CM, crust, and mantle along a great *circle* that connects their centers

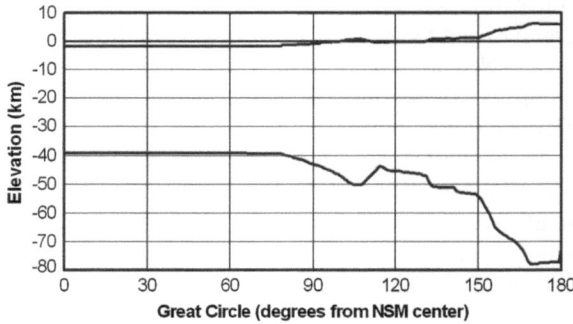

Fig. 8.7 Cumulative volume of ejecta from a vertical impact model (*circular*) of the CM as a function of the normalized internal radius of its apparent cavity. Recent GRAIL analyses of crustal thickness and porosity were not included in this figure (Wieczorek et al. 2012) or related text

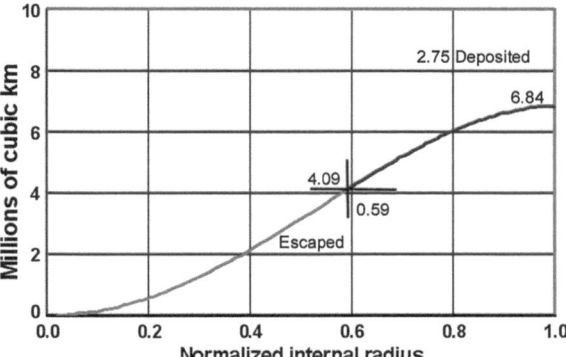

The total volume of material ejected from the original apparent crater would have been 6.84 million cubic kilometers (see Fig. 8.7). As the ejection cone expanded to a normalized radius of 0.59, 4.09 million cubic kilometers of ejecta escaped from the Moon. As the ejecta cone approached the rim, an additional 2.75 million cubic kilometers of ejecta would have been deposited on the rim and ejecta field.

8.3.4 The Floor and Melt Column of the CM

The CM melt column is inferred from the large size of the apparent crater and the flat floor, shown in Fig. 8.5. Future simulations should confirm that an impact of that size would form a melt column like that of the NSM and the SPA.

8.4 Rim and Ejecta Field of the CM

As mentioned, the western rim of the CM has interacted with the eastern rim of the NSM much like the interaction between the NSM and the SPA near the South Pole and like the interaction of the eastern rim of the Imbrium Basin with the western rim of the Serenitatis Basin. In all these cases, the earlier rim has been ejected by the edge of the later apparent crater and the later rim falls into the edge of the earlier apparent crater, leaving little evidence of any rim at all where they intersected. The effect is seen clearest in the case of the intersecting rims of the Imbrium Basin and the Serenitatis Basin.

The ejecta field of the CM underlies the Moscoviense, Mendeleev and Freudlich-Sharanov Basins. A review of Fig. 8.1 shows that the CM model does indeed improve the residual DEM in those areas.

8.5 Mineral Anomalies of the CM

The only mineral anomalies in the vicinity of the CM are contained within Mare Moscoviense. The Moscoviense Basin impact was in the area of the intersection of the NSM and CM rims, which, together with the shock waves from the two events, may well have weakened the crust there. The crust was also thinned by the Moscoviense Basin impact, followed by further thinning by the impact of a large crater within that basin, reducing the thickness of the crust there to essentially zero (see Fig. 7.9). These effects probably encouraged the rise of lava from the mantle.

The Clementine mission showed that Mare Moscoviense was rich in iron and titanium but the Lunar Prospector mission did not detect a thorium anomaly. This combination suggests that the incompatible layer may have been suppressed by the deep impacts into the weakened crust.

8.6 Age of the CM

Assuming that the CM followed the NSM and preceded the SPA, its age would be constrained by the time required to harden the CM and NSM melt columns. As discussed in Chap. 7, the ejecta from SPA fell on the NSM melt column before it hardened. The best fit model of the CM assumes that the ejecta from the SPA fell on the melt column of the CM after it hardened. This is possible because the time required for the NSM column to harden would be comparable to the time required for the lunar magma ocean to harden but the much smaller CM melt column would harden in much less time.

Taking the age of the NSM to be 4.34 Ga (from Chap. 5) and the maximum time to harden its melt column at 0.1 Ga (from Chap. 6) and the minimum time to harden the CM melt column to be 0.01 Ga, the age of the CM event, like that of the SPA event, would be between 4.34 and 4.24 Ga.

Like many estimates of absolute ages, these boundaries for the CM depend on a qualitative determination of sequence and quantitative estimations of intervals and absolute age of preceding and succeeding features. It is fortunate that we have samples from the NSM to provide a probable absolute age of that feature. We would probably do much better in constraining the age of the CM if we had samples from the SPA.

Chapter 9
History of the Crust of the Moon

9.1 Traditional History and New Paradigms

The history of the Moon is presented in the form of a series of topographic maps (DEMs) as the shape of the Moon is established. The history presented here follows the outline and content of "The Geologic History of the Moon" (Wilhelms 1987) with additions arising from concepts developed within the past decade. These include the algorithmic model of impact features developed in previous chapters of this book, the re-examination of the large craters and basins described in Chap. 4, and newly discovered features described in Chap. 5. These new features, the NSM and others, were found in the course of deconstructing the topography to create the comprehensive topographic model of Chap. 5.

In addition to the models developed here, two important external developments are woven into the history of this chapter. One is the new understanding of the melt columns of megabasins like the NSM, the SPA, and the CM, developed through advances in three-dimensional simulations. The other is the model of the development of the Solar System constructed by astronomers at Nice, France.

In this chapter, all the modeled features are reassembled in sequence, starting with the megabasins and proceeding through the stratigraphic units established by Gene Shoemaker and modified by Don Wilhelms (Wilhelms 1987). The geologic strata are called systems or series and those terms are adopted here for section titles. Within each section, other types of evidence besides superposition will be considered as well to estimate sequence, duration, and absolute age. The term period or epoch (a subdivision of a period) will be used to indicate time intervals instead of system or series but with the same distinctive names.

C. J. Byrne, *The Moon's Near Side Megabasin and Far Side Bulge*,
SpringerBriefs in Astronomy, DOI: 10.1007/978-1-4614-6949-0_9,
© Charles J. Byrne 2013

9.2 The Nice Model and Early and Late Heavy Bombardments

As the understanding of the time relationships of variations in the size and rate of projectiles advanced, it was suggested that there were two distinct periods when the size and frequency of new impact features were higher than the intervening and later periods. These two periods are called the Early Heavy Bombardment (EHB) and the Late Heavy Bombardment (LHB) (Tera et al. 1974; Taylor 2006).

A group of astronomers at the Côte d'Azur Observatory in Nice, France (Gomes et al. 2005; Tsiganis et al. 2005; Crida 2009) considered how the solar system evolved to the current arrangement of planets, moons, asteroids, and the Kuiper belt. They produced a new paradigm of formation called the Nice Model. According to this model, as the solar system evolved, the orbits of the giant gas planets became unstable and realigned. This phenomenon, taking place within an interval as short as 0.01 Ga, is hypothesized (Bottke 2012a, b) to have destabilized an E-belt of asteroids much more populous than today's asteroid belt. The E-belt is an extension of the main belt, discussed in Sect. 9.5. As a result, a cataclysmic dispersion of asteroids took place, resulting in the LHB of the Moon (as well as of Earth and other planets), leaving a new declining size and rate of impactors from the narrowed asteroid E-belt.

The EHB has been proposed as a period of a high rate of residual impactors from the formation of the planets, declining exponentially. As we shall see in the development of this chapter, the concept is replaced here with a short period of heavy bombardment followed by much lighter and uniform bombardment from asteroids in a broad primordial asteroid belt inferred in the Nice Model.

The impactors of the NSM, CM, and SPA megabasins discussed in Chaps. 5 through 7 appear to come from a separate population than the broad asteroid E-belt that caused the EHB. It has been suggested (Bottke 2012b) that they are the last of the planetesimals that formed Earth. The planetesimals are further discussed in Sect. 9.4.4.

Here, four distinct periods of bombardment, replacing the EHB and the LHB are described:

1. The three early megabasins were caused by the dwindling group of planetesimals that formed Earth.
2. A subsequent distribution of projectiles was from asteroids from the E-belt (extension from the main belt) inferred from the Nice Model (Bottke 2012a). This broad band of asteroids was fairly stable before the LHB so the size and rate of projectiles was diminishing but nearly constant.
3. The third set of projectiles resulted from a rearrangement of the gas giant planets that disrupted and greatly narrowed the early asteroid E-belt. This is the source of the LHB, also called the cataclysm.
4. The fourth set of projectiles comes from the narrowed band of asteroids left by the rearrangement of the gas giants. This is the population of impactors with a declining size distribution and frequency that extends to the current time.

In the course of describing the history of the Moon, these periods of early and late bombardment will be related to the stratigraphic layers.

9.3 How it All Began

The qualitative conventional view of the origin of the Moon is assumed here. A Mars-sized object approached the early Earth, struck it a glancing blow, and the impact vaporized most of the incoming body and part of Earth's crust, throwing them into an orbiting cloud of vapor, melt, and fragments. Most of the heavy core of that body became part of the Earth's core. The impacting body is sometimes named Theia, the Titan of Greek mythology who created light and became the mother of Helios (Sun), Selene, (Moon) and Eos (Dawn) (Halliday 2000).

The orbiting cloud rapidly condensed to form the Moon but lost much of its volatiles, those elements that are gasses in the space environment. The initial aggregated body, heated by the energy of the orbiting cloud and the conversion of potential energy to heat energy as it coalesced, included a magma ocean whose depth was at least 500 km and perhaps extended to the small core.

As it cooled, the magma ocean crystallized and the heavier minerals, mostly iron, magnesium, and calcium silicates, settled to form a basaltic mantle, and the lighter minerals, silicates of the lighter metals, rose to form the anorthosite crust. Between the crust and mantle, the minerals composed of elements that are slow to combine with others formed an incompatible layer of rare earth elements and concentrated uranium and other radioactive elements.

As the Moon solidified, it would have had a symmetrical ellipsoidal shape resulting from its rotational period. Assumption of this simple shape establishes a foundation for the subsequent history of its crust. Recent evidence from high resolution spectroscopic observations of exposed outcrops (Ohtake 2011) provides strong evidence that the separation of light minerals from the magma ocean was very efficient, resulting in a primarily homogenous anorthositic primitive crust. Even before the crust hardened, impacts may have left their signatures (see Sect. 9.4.4).

9.4 Early Megabasins and the Mounds: Pre-Nectarian Age Group 1

Once the magma ocean solidified, subsequent events were recorded in its crust. The earliest stratigraphic period, the pre-Nectarian, started with that solidification and, by definition, ended with the Nectaris Basin impact. The basins of the pre-Nectarian System have been classified into age groups 1 through 9, distinguished by successive overlap of the basins' craters and ejecta fields (Wilhelms 1987, Table 8.2). The earliest pre-Nectarian period, age group 1, is dominated by the three megabasins and the mounds discussed in Chap. 5. The shaping of the Moon by these three megabasins is shown in Fig. 9.1a–c.

Very soon after the magma ocean solidified, the Moon's dichotic shape was established. Although other hypotheses exist, the evidence presented here is in accord with the dichotic shape being caused by the first large hypervelocity impact after the crust hardened, the one that produced the Near Side Megabasin (NSM).

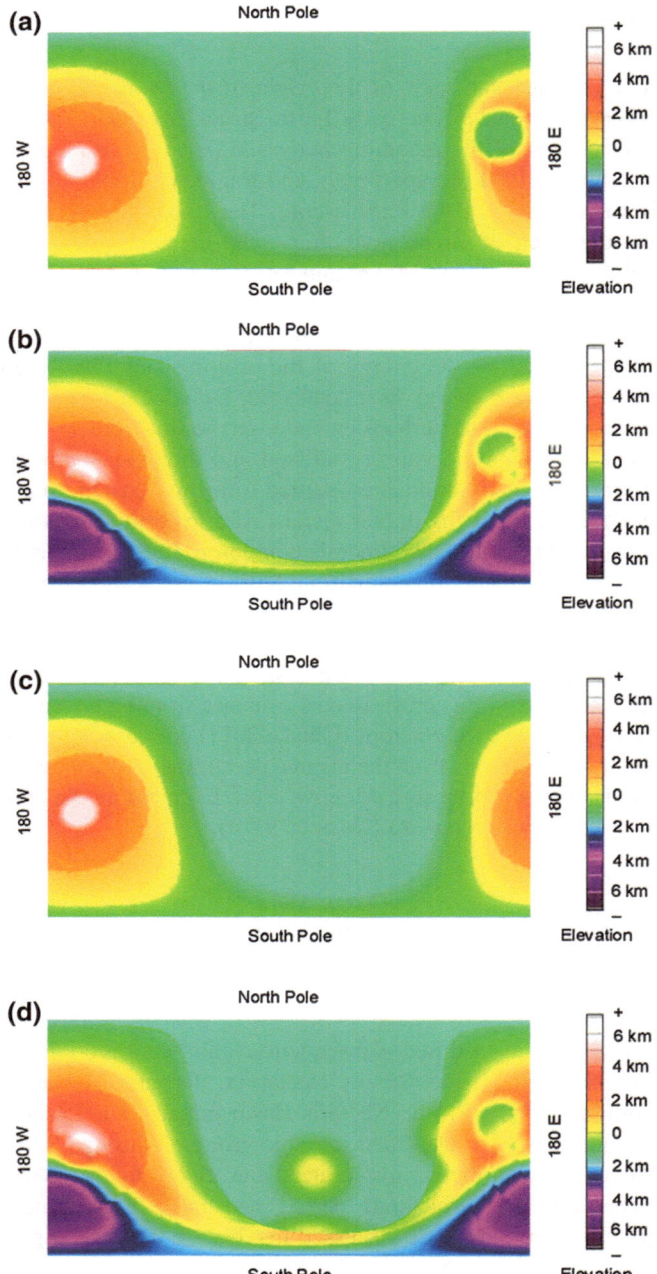

Fig. 9.1 The models of pre-Nectarisn age group 1, first NSM, then CM, then SPA and finally the four mounds. **a** Near side megabasin (*NSM*), **b** Chaplygin-Mandels'shtam Basin (*CM*), **c** South Pole-Aitken Basin (*SPA*), **d** Mounds

It struck what is now the near side, thinning the crust there. The resulting melt column penetrated the crust and far into the mantle, rose above the remaining crust as the melt and vapor expanded, and then collapsed, forming the crater's leveled floor. The ejecta from the NSM created the far side bulge. Two successive impacts (CM and SPA) modified the shape of the Moon while preserving its basic dichotic structure.

The South Pole-Aitken Basin (SPA) is clearly younger than the NSM because its cavity is free of the deposits of the far side bulge. The effect of isostatic compensation on the interacting NSM and SPA s is discussed in Chap. 7.

The Chaplygin-Mandel'shtam Basin (CM) has undergone full isostatic compensation (no free-air gravity anomaly) and is overlaid by other pre-Nectarian features (notably the Freundlich-Sharanov Basin). Superposition evidence alone can not establish the age sequence of the CM relative to the NSM and SPA. However, the best reduction in the standard deviation of the final residual DEM is obtained if the CM follows the NSM and precedes the SPA. That is the assumption for the model history here.

The three early megabasins share properties that establish them as unique in the history of the Moon.

- Large size
- Isostatic compensation indicated by free-air gravity
- Impact of their apparent crater by younger, smaller basins
- Ejecta of younger basins overlay the early megabasins.

Could later large basins such as the Imbrium Basin qualify as megabasins (Wilhelms 2012)? Yes, if their apparent diameter approaches or exceeds 1,000 km and they had other megabasin properties such as smaller basins within them or melt columns, perhaps found under mare flooding using gravity or sonic methods.

As the modeled features are added, the average elevation of each DEM varies as material escapes from the Moon, as the material deposited in the ejecta field expands in volume because it becomes more porous, and as melt column material rises, expanding due to heat and phase changes. The rising volume of material in the column is supplied initially by the heating and phase changes. Ultimately, as the column cools, it will be compensated by lateral movement of deep mantle material and general shrinkage as the entire mantle cools. Additional components of the average elevation budget will be supplied by the mounds of deposits, rise of lava to the maria, and depressions such as Vallis Procellarum.

The surface topographic variations of crater, rim, and ejecta field of the NSM and other early megabasins would have been much greater than today by a factor of about 6.0 because of the isostatic compensation that has taken place. Although that has reduced the surface imprint, the crustal thickness differences are preserved. Where the surface topography of the crater has become shallow, the lunar Moho (the boundary between crust and mantle) has risen to balance the void; where the rim and ejecta blanket have become less deep, the mantle has been depressed by the load.

9.4.1 The Lunar Surface After the NSM Impact

After the NSM impact, the model of the surface of the Moon would be as shown in Fig. 9.1a. The boundary of the NSM apparent crater is at the green-yellow interface. The rim and ejecta field of the NSM rise from yellow through orange and red to white. The blue area is the flat floor, the top of the NSM melt column. The elevations here are after isostatic compensation. The value of the elevation of the initial crust before the NSM impact has been found by optimization of the final model to be −320 m below the area-averaged elevation of the Kaguya 1° DEM, −240 m.

The estimated absolute age of the NSM is 4.34 Ga, based on the plurality of zircon grains in thin samples of rocks collected at the Apollo 14 site in the ejecta from the Imbrium Basin and the majority of zircon grains in rocks collected at the Apollo 17 site inside the rim of the Serenitatis Basin, as discussed in Chap. 6. This is an important reference point in the absolute ages of the Moon's features.

9.4.2 The Model After the CM Impact

Figure 9.1b shows the model after the projectile of the CM megabasin has impacted, striking the northwestern slope of the far side bulge. As the SPA will do later, the CM deposits ejecta onto the peak and side slopes of the far side bulge beyond its apparent crater. The CM melt column has depressed and leveled the surface.

The model of the CM calls for ejecta to be deposited on the floor of the NSM, but there is no evidence of this in the current topography. It is likely that the NSM flat floor was still molten or highly plastic at the time of the CM impact so the model restores the flat floor of the NSM.

9.4.3 The Model After the SPA Impact

Figure 9.1c shows the model after the impact of the SPA on the southern slope of the far side bulge. The deep impact of the SPA on the far side bulge has also spread its ejecta on the top of the far side bulge and onto the CM.

In addition to the obvious deposits of ejecta on the far side bulge and the flat floor of the CM, there should have been ejecta on the flat floor of the NSM (including an antipode deposit), but there is no evidence of this on the current surface. This could either be due to the surface of the NSM melt column still being molten or very plastic as the SPA ejecta was deposited as discussed for the CM or it could be due to the SPA ejecta field being distorted by the oblique impact. Here, the flat floor of the NSM is assumed to be molten at the time of the SPA impact and the model restores the NSM flat floor. That assumption constrains the interval between the NSM and SPA events to less that 0.1 Ga, based on an upper bound of 0.2 Ga for the time for the LMO to crystallize (Shearer 2006).

9.4.4 The Planetesimals: Cause of the Early Megabasins

The largest of the features in the pre-Nectarian System that followed the megabasins and the mounds is Nubium with an apparent diameter of 712 km while the diameter of the CM, the smallest of the earlier megabasins, is 1,320 km. This suggests that the megabasin projectiles and the projectiles of the rest of the pre-Nectarian impact features are members of two different populations. In the early solar system, dust grains adhered as they interacted, until, by chance, bodies grew to the size of 1 km or greater, at which point they had enough self-gravity to hold together. Beyond that size they are termed planetesimals. Further interaction results in growth of the planetesimals and further aggregation into the planets. The source of the projectiles that produced the megabasins could have been the last of the planetesimals that were attracted by Earth's gravity (Bottke 2012b).

Three of these planetesimal impactors have been identified as the cause of the early megabasins. Based on area ratios, there would have been about 51 planetesimals impacting Earth at this time. There are two other features that have been identified as basins in the past but have not been confirmed as impact basins because there is no clear topographiic rim or ejecta field. These could be caused by planetesimals that impacted before the crust hardened. One is the Procellarum impact feature (Nakamura et al. 2012) and the other is the Australe impact feature (Byrne 2013). Both have been identified by mineral paterns associated wieh their melt columns.

9.4.5 The Modeled Mounds of Pre-Nectarian Age Group 1

As described in Chap. 5, examination of the residual DEM after modeling the early megabasins and the other large craters and basins revealed four positive features that were modeled as circular mounds with a raised-cosine profile. Mons Nectaris, 1,900 km high and 2,120 km in diameter, models the northernmost part of the near side central highlands; its eastern slope has been impacted by the Nectaris Basin. Another mound is on the near side near the South Pole, one is southwest of the CM megabasin, and there is a shallow mound northeast of that megabasin, past the 180° meridian.

The mounds are attributed here to return of ejecta from the early megabasins that had escaped the Moon (but not the Earth-Moon system) and then returned to the Moon at near the lunar escape velocity. An alternate explanation for these mounds is that they were formed by uplift due to a pluton that did not break through the crust. This seems unlikely in the pre-Nectarian period. The Moon's topography after the addition of models of these four mounds is shown in Fig. 9.1d.

These mounds are not associated with gravity anomalies on the free-air gravity map, so it is assumed that they were deposited in an early age group. Further modeling and examination of residual DEMs shows no evidence of any other modeled features than the early megabasins having preceded the mounds. Therefore, they are assigned here to age group 1 of the pre-Nectarian period.

Since Mons Nectaris was not absorbed, the flat floor of the NSM must have hardened by the time it was deposited. Therefore, this and the other mounds are assigned here to the end of pre-Nectarian age group 1, defining the boundary between groups 1 and 2. Of course that does not bound the time between the first of these four mounds and the last nor does it establish their sequence; it simply assumes that they are prior to any basins currently assigned to Groups 2–9. It is possible that future basins or other features may be found and assigned to age group 1 despite the difficulties in establishing relative ages between such features, the mounds, and the megabasins.

9.5 The Large Craters and Basins of the Pre-Nectarian System (Age Groups 2 Through 9)

Without further separation between age groups of the pre-Nectarian system but retaining the sequence described in (Wilhelms 1987, Table 8.2), the modeled lunar topography after the end of that period is shown in Fig. 9.2a. The craters, rims, and ejecta fields of these features are superposed on the topography established by the early megabasins and mounds.

Since the sequence of the models is realistic, much of the interplay between pre-Nectarian features is captured. For example, the impact of the Smythii Basin on the Marginis Basin is shown. Also, note the impact of the Apollo Basin and Ingenii Basin on the SPA. At this time, the craters and basins have partial fill from the ejecta from other features, but no maria, which came later.

As mentioned above, the Nubium Basin (712 km) is modeled as the largest apparent diameter in pre-Nectarian age groups 2 through 9, only slightly more than half as large as the CM. Clearly, the projectiles of the pre-Nectarian groups 2 through 9 come from a different source than the planetesimals drawn to Earth. According to the Nice Model, that source would have been the E-belt that extended between 1.6 and 2.1 AU from the Sun (Bottke 2012a). Its population would be asteroids, collision fragments and small planetesimals that never coalesced into a planet. The incoming velocity to the Moon would vary widely according to the specific source radius and the position of the Moon in its orbit at the time of encounter (and therefore the relative velocity). A simulation finds a median velocity of 8.6 km for this population of projectiles (Bottke 2012a, Fig. S1). The early megabasin projectiles, planetesimals formed near 1 AU, would have had a lower velocity as they impacted the Moon, but a much larger mass.

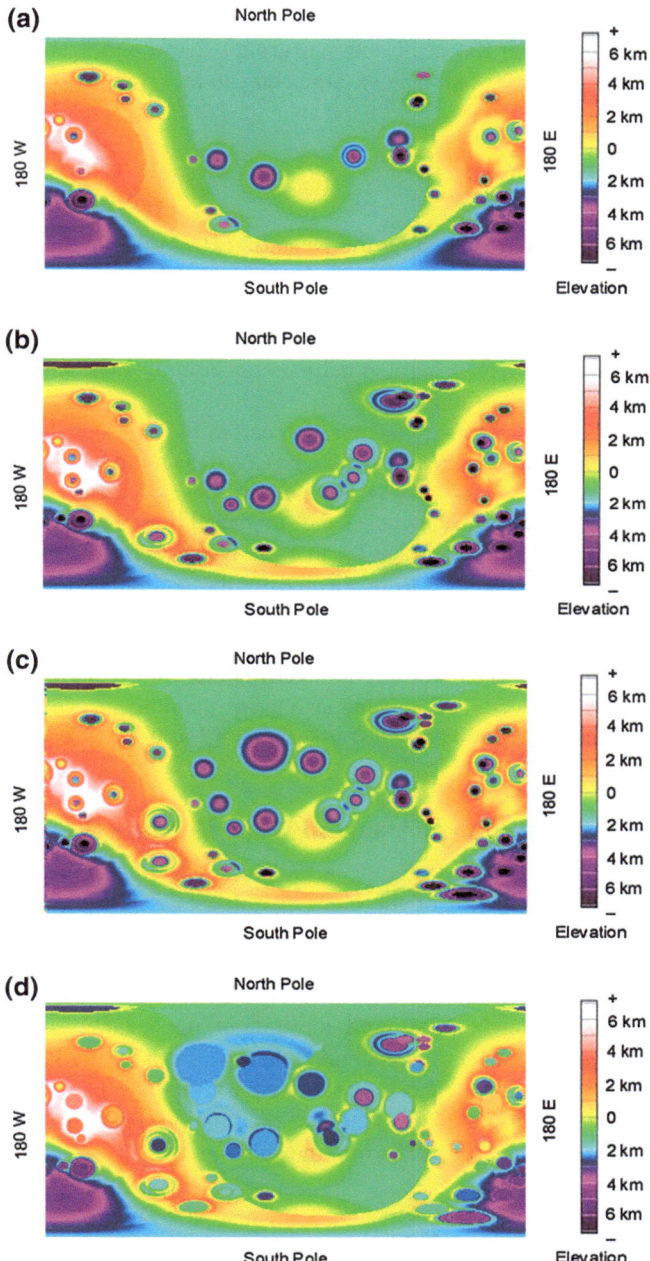

Fig. 9.2 The models after each of the indicated stratigraphic periods. **a** Pre-Nectarian period, **b** Nectarian period, **c** Early Imbrium epoch, **d** Late Imbrium epoch

It is not impossible that small planetesimals near 1 AU could have caused some of the impact features of the pre-Nectarian groups 2 through 9 but the available population in the ancient E-belt is very large and can be expected to dominate.

9.6 The Nectarian System

The Nectarian period is bounded by the Nectarian Basin impact and the Imbrium Basin impact. It is a time of increasingly large impact features, starting with the Nectaris Basin (840 km) itself. The model after the Nectarian period is shown in Fig. 9.2b. The Nectarian period is arranged in two age groups, and basins are sequenced within each age group, sometimes uncertainly (Wilhelms 1987). The models are sequenced accordingly.

The circular mound that formed the northern part of the near side central highlands has been obscured by the Nectaris Basin and the CM megabasin has been obscured by the Moscoviense and Mendeleev Basins. That is why these features were not identified until the obscuring features were modeled and subtracted from the current topography.

It has been suggested (Bottke 2012a) that the Nectaris Basin's projectile is the first of the many E-band bodies to be ejected by the realignment of the giant gas planets. The impact features of the Nectarian period may be a mix of late arrivals from asteroids ejected from the E-belt before the LHB and early arrivals from those ejected during the LHB. The Crisium Basin (740 km, Nectarian age group 2), almost as large as the Nectaris Basin and also larger than the pre-Nectarian Nubium Basin is also a candidate as an early disturbed asteroid of the LHB.

9.7 The Disturbed E-Belt Asteroids that Became the LHB

The effect of the realignment of the giant gas planets is that the ancient E-belt, ranging 1.6–2.1 AU, was trimmed by ejecting a large population of asteroids at the low and high ends of the range. It also ejected a population with low inclination across the entire range (Bottke 2012a, Fig. 9.1).

These disturbed asteroids scattered throughout the solar system. No doubt many arrived at the Sun. Many others ended their trajectories at planets, and some at moons, including our own. Those that arrived at the Moon had a median velocity 20.7 km/s (Bottke 2012a, Fig. S1), faster than the asteroids from the primeval belt because they came from further from the sun.

9.8 The Lower Imbrian Series

The Early Imbrian epoch starts with the Imbrium Basin impact and ends with the Orientale Basin impact, succeeded by the Late Imbrium epoch.

The Imbrium Basin continues a pattern of larger basins started with the Nectaris Basin and continuing with the Crisium, Serenitatis, Imbrium, and

Orientale Basins. The models of the Early Imbrian period are shown in Fig. 9.2c. The massive ejecta blanket of the Imbrium Basin is seen here as turning part of the blue false color of the NSM crust (see Fig. 9.1c) to the higher elevation shown as green. The interaction between the Imbrium and Serenitatis rims is also shown. They have mutually obliterated each other where they overlap and where they just approach each other they have raised the elevation. Similar effects can be seen between other basins that are near each other.

Clearly, in the Early Imbrian epoch, many of the basins are larger than previous features. Most or all of the projectiles that formed these large basins have been disturbed from the E-belt asteroids.

9.9 The Upper Imbrian Series

The Late Imbrian epoch, starting after the Orientale Basin impact, was extraordinary; not only because the LHB continued and terminated in that period but because the major maria were established (see Fig. 9.2d. The end of the late Imbrian epoch is established not by an impact event but by the deposit of specific layers of lava. This also marks the beginning of the Eratosthenian period and is discussed further in Sect. 9.10. Creation of the major maria and establishing the level of fill of the associated basins are the most striking phenomena of this period, along with the establishment of the Vallis Procellarum depression.

Subsequent eruptions added layers of lava into the Eratosthenian and Copernican periods but few new maria were added after the Late Imbrian period. The Vallis Procellarum depression could have occurred in this period as the depletion of a massive pool of lava, together with the weight of the erupted lava above the crust allowed the subsidence of the crust in that area. This may have stimulated still more lava eruptions.

The initiation of lava flows in this period is due to the accumulated heat of radioactivity in the incompatible layer and upper mantle, as transformed by the NSM hundreds of millions of years earlier. Isn't it ironic that this interval of massive volcanism could be associated with a much earlier impact? A further contribution of impact to volcanism is the last of the large impacts of the LHB that formed the Lavoisier-Mairan Basin (just west of the Imbrium Basin). Like other basins in or near Vallis Procellarum (Nubium, Humorum, Flamsteed-Billy, Cardanus-Herodotus, and Imbrium), its probable melt column not only thinned but penetrated the crust, contributing its heat to the pool of lava formed or forming there.

The first period of large impact features was the pre-Nectarian age group 1, the megabasins. The second period of large impact features, the LHB, began in the Nectarian period and ends in the Late Imbrian epoch. The largest impact feature after this period is the Eratosthenian Hausen crater, only 148 km in diameter.

Between the Lavoisier-Mairan and Imbrium Basins is an interesting area containing the Aristarchus plateau, large lava flows such as Vallis Schroterii, and pyroclastic flows. If there is an epicenter of lava eruption, Vallis Procellarum melt pool depletion, and crustal subsidence, it is here.

Knowing that the LHB ends in this period, it is interesting to look at the whole phenomenon, probably starting in the Nectarian period, building up in the Early Imbrium epoch and finally ending in the Late Imbrium epoch. The five largest modeled basins since the time of the planetesimals (Imbrium, Lavoisier-Mairan, Orientale, Nectaris, and Crisium basins) struck in these three periods, all within the rim of the NSM (all but the Orientale basin are within the flat floor, the top of the melt column, of the NSM).

One wonders if these largest LHB basins left the E-belt from the same areas, followed similar trajectories, approached but missed Earth, were focused by Earth's gravity (Byrne 2005), and struck the Moon when it happened to be in the right place at the right time. These impacts need not have been nearly simultaneous; the links among them could be the manner of their dispersion from the E-belt of asteroids and their subsequent trajectories. This suggestion is not about an improbable salvo but it provides a speculative answer to the question "How do you make a really large lunar basin during the LHB?"

9.10   The Eratosthenian and Copernican Systems

Lava continues to flow in the Eratosthenian period but it mostly covers or extends areas that were established in the Late Imbrian epoch. The start of the age period of the Eratosthenian System is defined by the average estimated age (3.2 Ga) of the mare basalt sampled at the Apollo 12 (3.16 Ga) and Apollo 15 (3.26 Ga) landing sites (Wilhelms 1987). The crater Eratosthenes is the type example as Copernicus is the type example of the later Copernican period (see Fig. 9.3). Both have a fresh appearance relative to craters of the earlier periods, a consequence of the termination of the LHB in the previous period (Fig. 9.3).

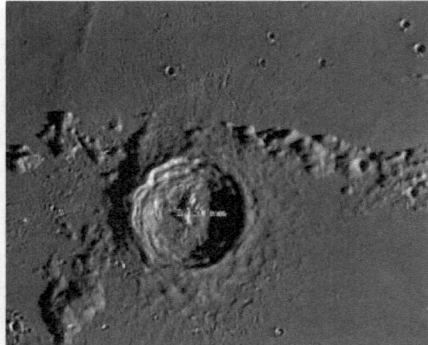

Fig. 9.3 The craters Eratosthenes (*right*, 58 km) and Copernicus (*left*, 81 km) are the type examples of the Eratostenes and Copernican periods. *Eratosthenes source* Wikimedia, David Campell. *Copernicus source* Winer Observatory, Mike Shade

The outstanding characteristic of Eratosthenes that distinguishes it from Copernicus is a lack of the bright rays that Copernicus displays across both mare basalt and crustal anorthosite (Hawke 2007). Rays fade with time as they are darkened with exposure to the solar wind and (relatively slowly) as the brighter crustal minerals are gardened into a basaltic target surface.

The Copernican period extends to the current time. Astronaut's footprints could be considered as Copernican features as well as manned and unmanned landers. Several new Copernican craters have been caused by the accidental impact of spacecraft components or deliberate impacts to keep inoperative units from being hazards to future missions. In the case of the Lunar Crater Observation and Sensing Satellite (LCROSS) a projectile and an observation instrument were used to measure concentrations of volatiles, including water, in permanently shadowed areas. Perhaps in the future a new geologic period will be established: the age of human activities.

9.11 Absolute Ages

What is history without dates? While there are many estimates for the absolute ages of lunar features and for the duration of the major geologic periods, there is only limited consensus.

The accretion of the Moon is estimated to be, at the latest, 4.52 Ga (Lee et al. 1996). The hardening of the lunar crust can be taken as the beginning of the pre-Nectarian System. The age for the solidification of the magma ocean has been estimated by two different methods. The age of zircon crystals that formed as the incompatible layer crystalized has been measured in a sample returned from the Apollo 17 landing site as 4.417 ± 0.006 Ga (Pidgeon 2010). Argon–Argon dating of anorthosite samples from the Apollo 16 and 17 landing sites has been used to measure the crystallization age of the crust as 4.44 ± 0.02 Ga, 4.43 ± 0.05, and 4.549 ± 0.054 Ga (Fernandez et al. 2008). Taking these measurements together,

Table 9.1 Estimated absolute ages

Event or interval	Beginning Ga	End Ga
Moon accretion		4.52
Crust hardened		4.40
Pre-Nectarian		
Age group 1	4.40	4.20
Age group 2–9	4.20	3.92
Nectarian	3.92	3.85
Early Imbrian	3.85	3.75
Late Imbrian	3.75	3.20
Eratosthenian	3.20	0.80
Copernican	0.80	Current time

the crust would have hardened, at least in most areas, by an estimated 4.4 Ga. This is also the oldest age for the hardening of at least parts of the crust in "The Geologic History of the Moon" (Wilhelms 1987).

Estimates of the order of 0.1 Ga for conductive cooling of the lunar magma ocean are consistent with the 4.4 Ga estimate for the age of the hardening of the lunar crust, the beginning of pre-Nectarian age group 1.

The estimated duration of pre-Nectarian age group 1 here is based on the sequence of the three megabasins and interactions among them. Analysis of zircon grains in samples from Apollo 14 and 17 indicate an age of the NSM to be 4.34 Ga (see Chap. 6). The NSM melt column would be still molten at the time of impact of the SPA (see Chap. 7 and Sect. 9.4.3). Additional time would be required for the mound deposits (see Sect. 9.4.5). The end of age group 1 is assigned here to be about 4.2 Ga, making its duration to be 0.2 Ga. It is not likely to be much more, but could be much less, depending on the length of time for the melt column of the NSM to harden and the mounds to form.

Table 9.1 shows the list of ages used in this history. Ages after pre-Nectarian age group 1 were compiled from estimates summarized in detail in "New Views of the Moon" (Stoffler et al. 2006, Fig. 5.31) with reference to diverse sources (e.g., Wilhelms 1987; Stoffler and Ryder 2001). In taking ages from (Stoffler et al. 2006, Fig. 5.31), the age of the Imbrium Basin as taken to be 8.5 Ga (option b of the reference).

Chapter 10
Methods, Results, and Future Directions

10.1 Methods

The methods used to complete the model of the Moon's large scale features are reviewed with a concise statement of the results. In the course of this work, new tools were used to model the large features of the Moon. The history of the Moon was investigated by sequencing the formation of those features to establish relative age and apply what age information is available to estimate absolute ages where possible.

10.1.1 New Paradigms

Three relatively new paradigms have influenced this account of the shape and history of the Moon:

- The dynamic nature of the melt columns central to large impact features, discovered by three dimensional simulations of large impacts
- The Nice model of disruption of the asteroid belt that caused the LHB
- The three early megabasins are a distinct group of impacts resulting from planetesimals that formed near Earth's orbit.

   The melt columns are related to aspects of lunar thermal history that underlay age estimates of the early megabasins. They explain the flat floors of large impact features. The melt column of the NSM turbulently mixed materials from the mantle, crust, and incompatible layer, varied the mineral compositions and set the stage for nearly all mare flows. All returned samples from the Moon come from the current surface of the melt column of the NSM.

   The Nice model of the solar system explains the LHB. The prior knowledge of the lunar LHB helped to establish the model of the entire solar system. The model developed at Nice also identifies the source of the impacts between the megabasins

C. J. Byrne, *The Moon's Near Side Megabasin and Far Side Bulge*,
SpringerBriefs in Astronomy, DOI: 10.1007/978-1-4614-6949-0_10,
© Charles J. Byrne 2013

and the LHB as coming from the broad primordial asteroid belt. It also explains the sudden transition from the LHB to the progressively lower rate of impacts in the Eratosthenian and Copernican periods.

The concept of the Early Lunar Bombardment has been that an initially high rate of planetesimals decreased in the pre-Nectarian period. Instead, this is seen here as consisting of two populations of impacts. One was the three megabasins, the last of the planetesimals. Another group of early impacts resulted from fall-out from a broad primordial belt of asteroids.

The LHB is seen as a relatively short period of very heavy bombardment due to the realignment of the giant gas planets having ejected the majority of the asteroids in the E-belt, greatly narrowing that part of the primordial belt. After the LHB the rate and size of the impacts declined.

Identifying two additional megabsins, the NSM and the CM to accompany the SPA completes the explanation of the general shape of the Moon. The inference that the projectiles of the megabasins were planetesimals suggests that additional planetesimals may have struck the Moon before the lunar magma ocean hardened. About 17 times as many planetesimals of the same size would have struck Earth in this period, with implications to its early history.

10.1.2 Modeling the Features of the Moon's Crust

A comprehensive model of the shape of the Moon has been assembled through superposition of 3 megabasins, 33 basins, 16 craters with apparent diameters of 200 km or more, 2 depressions, and 4 mounds. These are all the features of their size and type that are identifiable in the final residual DEM and can be modeled. In addition, 17 craters with apparent diameters of less than 200 km have been modeled. The megabasins, basins, and craters are all modeled with a single algorithm for hypervelocity impact features.

The impact feature algorithm was based on a combination of several extensions of the Maxwell-Z Model. This extended model describes the apparent crater, rim, and ejecta field of all lunar impact features, given their apparent diameter, apparent depth, and fill. These parameters are determined for each feature by estimating and removing the pre-impact shape of the target surface.

The features that fundamentally shaped the Moon were found to be three megabasins, the familiar South Pole Aitken Basin, the previously identified Near Side Megabasin, and a newly identified megabasin, the Chaplygin-Mandel'shtam Basin. Additional new features include the Vallis Procellarum, the arc shaped depression underlying Oceanus Procellarum and the identification of the northern near side central highlands as a large circular mound that was subsequently impacted by the projectile that formed the Nectaris Basin.

All of the traditionally identified basins (Wilhelms 1987) have been examined with the new modeling method. Models of many were determined, with some adjustment in the traditional parameters, especially in revised identification of the main ring in the case of multi-ringed basins. Models could not be formed for some of the traditional basins, especially those identified as "probable" or "possible". Inability to model the features by the method used here does not necessarily establish that it is not an impact feature: it may be too degraded or flooded to be modeled by the methods used here. Models were formed of several basins reported by other investigators and additional new features identified by examination of intermediate residual DEMs.

10.1.3 Sequencing the Modeled Features

The history of the lunar topography, crust, and upper mantle has been inferred from the large features that were identified from the current topography. The inferences are based on three dimensional analytic models of hypervelocity impact features and a few additional features that are identified by examination of the residual DEMs constructed by subtracting the growing set of impact models from the current topography.

The melt columns of the largest modeled impact features penetrated deep into the mantle. This has been known but there were few signs of mantle material being excavated because the melt columns collapsed on themselves and differentiated (Stewart 2011).

Excavation of apparent cavities and deposit of ejecta onto rims and ejecta fields are identified as the primary source of crustal thickness variations. Isostatic compensation of the largest and oldest features reshapes the lunar Moho, the boundary between crust and mantle.

These features have been placed in historical sequence based on their causal relationships derived from stratigraphy and a few absolute ages determined from analysis of lunar samples returned from the Apollo and Luna programs, as published in the references.

10.2 Results

10.2.1 The Shape of the Moon

The Maxwell Z-model has been extended, with the aid of double normalization (by diameter and depth separately), to derive an algorithmic function for the apparent crater, rim, and ejecta fields of all circular hypervelocity impact features from laboratory scale up to the NSM, whose apparent crater covers more than half of the Moon.

The dichotic shape of the Moon is due to the three early megabasins. The comprehensive model reduces the variance of the DEM of the current topography by 76 %. Of that reduction, 89 % is attributable to the three megabasins. These impacts also accounted for the dichotic thickness of the crust and the other major asymmetries of the Moon. In the course of the modeling work, several new features were identified and have found their place in the history of the Moon.

A conclusion that can be drawn is that the shape of the Moon shortly after the crust hardened was essentially an ellipsoid with a uniform crustal thickness. Three hypervelocity impacts shortly after the crust hardened, caused by planetesimals drawn toward the Moon by Earth's gravity, caused the major variations in the Moon's shape.

The spatial variations in mineral composition of Apollo samples could have been caused by turbulence in the NSM melt column; convection in the Lunar Magma Ocean was not required.

10.2.2 The History of the Moon

Absolute ages have been estimated or constrained for the lunar features, in pre-Nectarian age group 1. This age group originally included the SPA and the Procellarum Basin. Here, the NSM and CM have been included in Age Group 1, along with four large mounds, with relationships among these features used to delimit the period of pre-Nectarian age group 1. The Procellarum impact feature, lacking a rim and ejecta field, is likely to have preceded the hardening of the crust and therefore preceded pre-Nectarian age group 1.

Sequence and age estimates for stratigraphic periods later than pre-Nectarian Age Group 1 are taken from "The New Views of the Moon", Part 5, Cratering History and Lunar Chronology (Stoffler 2006, Sect. 6) or from "The Geologic History of the Moon" (Wilhelms 1987). Several other new features discovered in the modeling process have been placed in an estimated sequence with the traditionally known features.

The LHB began in the Nectarian period and continued in the Lower Imbrium period; most of the large basins were formed in these two periods. The maria and mare sheets between them were primarily established in the Upper Imbrian period, during the end of the LHB. Most mare flows in later periods were limited to surface flows of new layers of lava over Upper Imbrian flows. The last of the large basins were formed in this interval. Apparently there were two factors coincidentally converging in this time interval: formation of a large melt pool in the incompatible layer and upper mantle and the roughly simultaneous formation of the last basins of the heavy bombardment.

The melt columns of these large impact features not only thinned the crust but also penetrated it into the melt pool and added heat. The combination apparently caused release of the melt pool through and over the crust, depressing the shocked and fractured crust into the depleted melt pool. The resulting depression is modeled as an arc-shaped depression and called here Vallis Procellarum.

It has been common to justify study of the Moon as a witness plate. A witness plate is a passive test device that records surrounding events on its surface.

The Moon has been a relatively stable body that records a sample of the history of the Solar System on and in its crust, with minimal degradation over time. It has been such an informant since Galileo saw through his early telescope that it was a very irregular cratered body, not an ideal spherical form. The concept has been applied here, with the help of the recent new paradigms, to relate the history of the Moon to the history of the Earth and the entire Solar System.

Interpreting the information recorded on the witness plate has not been easy. Lunar Orbiter's high resolution and comprehensive medium resolution photography was complete in 1967 and the Apollo orbital photography, Apollo samples, and Luna samples and were returned to Earth a few years later. Yet it took until 1984 to achieve a consensus on the Moon's origin.

Analysis of the ages of rock samples showed a concentration of ages around 3.8–3.9 Ga, implying the LHB but it took until 2005 to understand that the LHB affected all the rocky bodies of the solar system and that it was due to the realignment of the giant gas planets.

Here, it is proposed that the record of the last planetesimals to complete the aggregation of Earth was written large in the early lunar megabasins. The underlying data is based on topographic evidence collected by Clementine in 1991 after hints in the Apollo missions and before improvement by Kaguya and currently the Lunar Reconnaissance Orbiter.

Altogether, the Moon has taught us that the solar system was a complex place and that our comfortable Earth today is the result of a remarkable series of chaotic processes. Yet we can model and understand statistical properties of those processes to prepare ourselves to study the new solar systems being discovered in our galaxy.

10.3   Future Directions

A straightforward advance in this work would be to continue to model the smaller features, perhaps down to craters 100 km or smaller. Subtracting the models of such feature and examining the new residual DEM would possibly reveal new features.

The model of the topography and crustal thickness, together with estimates of the density of crust, mantle, and core, could be used to determine moments of inertia and offset of the center of gravity from the center of mass and compare the result with observations. The model could also be used to derive an estimate of free air gravity anomaly for comparison with the GRAIL measurements.

Three dimensional hydrocode simulations of the South Pole-Aitken Basin and the Orientale Basin are available (Stewart 2011, 2012; Andrews-Hanna 2011). It is very desirable to provide such simulations for the NSM, the CM, and a series of basins of intermediate size such as the Imbrium Basin. Such a series would shed light on the relationships among diameter, depth, and level of crustal fill. It would then be possible to estimate the degree of isostatic compensation of individual lunar craters and refine estimates of the depth of mare in those basins that have been flooded. Further, the ability to infer a melt column under a covering maria may be helpful in the interpretation of gravity anomalies.

A further advantage of a comprehensive exploration by simulations would be refinement of the contribution of melt columns to the thermal history of the Moon. Studies of the chaotic mix of crust and mantle materials in these melt columns may lead to an improved understanding of the mineral variations previously attributed to mantle convection.

Returned samples from the far side of the Moon would provide new information relevant to the ages of the major features there, refining estimates of pre-Nectarian Age Group 1 timing. Seismic data from the far side of the Moon would determine whether the depth of the active layer measured on the near side (500–600 km) extends to the far side of the Moon or is associated with melt columns The result would be relevant to the depth of the Lunar Magma Ocean and to the depth of the megabasin melt columns.

Errata to: The Moon's Near Side Megabasin and Far Side Bulge

Charles J. Byrne

Errata to:
C. J. Byrne, *The Moon's Near Side Megabasin and Far Side Bulge*, **SpringerBriefs in Astronomy,**
DOI 10.1007/978-1-4614-6949-0

Underlined text is the corrected text. Hence, the sentences should read as given in "Text should read" column in the below tables.

The online version of the original book can be found under DOI 10.1007/978-1-4614-6949-0

C. J. Byrne (✉)
Middletown, NJ, USA
e-mail: charles.byrne@verizon.net

C. J. Byrne, *The Moon's Near Side Megabasin and Far Side Bulge,*
SpringerBriefs in Astronomy, DOI: 10.1007/978-1-4614-6949-0_11,
© Charles J. Byrne 2013

Errata

Page number	Location	Text should read	Remarks
8	Section 2.2, first paragraph	The term "basin" was introduced by <u>lunar researchers including those from</u> the United States Geologic Service (USGS).	
16	Section 3.1.1, first paragraph	A new method …digital elevation maps … Lunar Reconnaissance Orbiter (<u>an example is shown in Fig. 1.2</u>).	
54	Last paragraph	3,950 <u>m</u>	*Note*: The error occurs twice in this paragraph.
65	Third paragraph	My response … double <u>the length and width of</u> the area vetted for safety.	
70	Fig. 6.16 legend	…(higher <u>lunar Moho, the lunar equivalent of Earth's Mohorovičić discontinuity</u>) than is shown here.	
71	Section 6.4.1, next to last paragraph	…(typically 10–30 μ<u>m</u>)…	
78	Section 7.3, second paragraph	…. Depth of the <u>SPA</u> crater…	
85	Top	In a vertical impact: … that follows the <u>geoid</u>, the material …	
90	Fig. 8.2		*Note*: Artwork color should agree with legend (see figure corrections below).
100	Fig. 9.1		*Note*: The order of maps should be corrected (see figure corrections below).
104	Section 9.5, third paragraph, near bottom	A simulation finds a median velocity of 8.6 <u>km/s</u> for this population …	
108	Second paragraph	…. largest LHB <u>impactors</u> left the E-belt …	
110	Last sentence	… the age of the Imbrium Basin <u>is</u> taken to be <u>3.85</u> Ga …	
115	Section 10.3, second paragraph	The model … offset of the center of gravity from the center of <u>figure</u> and compare …	

Typos and Clarity Corrections

Page number	Location	Text should read	Remarks
33	First paragraph	The full set ...radial profiles of the set...	
49	First line	The Near Side Megabasin (NSM) is ...of its apparent crater that covered...	
62	Section 6.2.6, last paragraph	There is an irony ... impact or volcanism ...	
66	Section 6.2.9, second paragraph	Vallis Procellarum	
69	Section 6.3.4	The GRAIL mission ...lowest altitude of its mission...	
70	Section 6.3.4: last paragraph, last sentence	...it would be a constraint...	
71	Section 6.4.1, last paragraph, first sentence	An improved measuring instrument (Kennedy and de Laeter 1994) was used to analyze zircons from several samples from two widely separated Apollo landing sites. Precise ages were determined for several events that have been strong enough to reset the zircon clocks (Nemchin et al. 2008, 2009).	*Note*: The first sentence should be replaced with the given text.
74	Section 6.5, next to last paragraph	...Kaguya spectroscopy...	
75	First and second paragraphs		*Note*: "Identification of the SPA ...southern part of its cavity" repeated twice. Duplicate should be deleted.
80	Section 7.3.3, fourth paragraph	The maps in ...from Kaguya data, (Sasaki et al. 2011) that present a coherent	*Note*: Change period to comma after the word "data", and underlined word should be "present" not "presents".

(continued)

(continued)

Page number	Location	Text should read	Remarks
83	Section 7.3.4, second paragraph	...one of S. T. Stewart's simulations are...	
83	Last line	... was ejected vertically collapsed into the crater.	
84	Fig. 7.10 legend	... 3-D simulation of an SPA impact ...	
87	Section 7.6, end of first paragraph	So the SPA is probably younger than the NSM, which was ...	*Note*: "(SPA)" after the term "NSM" should be deleted.
99	Section 9.3, second paragraph	...those elements that form gasses in the space environment.	
101	Second paragraph	... the interacting NSM and SPA is discussed in Chap. 7.	*Note*: The letter "s" should be deleted after the term "SPA".
103	Top	...for the Lunar Magma Ocean (LMO) to crystallize ...	
103	Section 9.4.4, last sentence	... mineral patterns associated with their melt columns.	
112	Third paragraph	Identifying two additional megabasins, the NSM and the CM, to	*Note*: Underlined word spelling has been corrected, and comma has been inserted after "CM"
115	Third paragraph	Analysis of the ages ... 3.8–3.9 Ga, implying the LHB, but ...	*Note*: Comma should be added after the term "LHB".
116	Next to last sentence	... is associated with melt columns.	*Note*: End period should be added.

Revised Figures

Figure 8.2 should be:

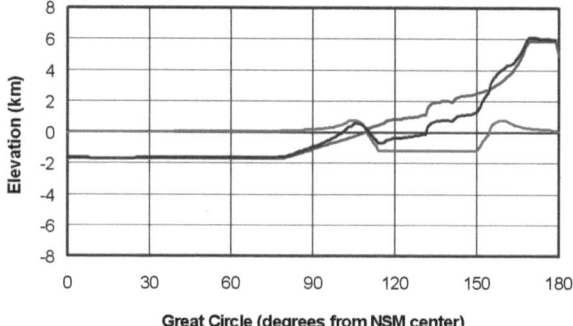

Fig. 8.2 The interaction between the current topography of the CM and the NSM is shown along a great *circle* that connects the centers of the two megabasins with the angle proceeding toward Moscoviense. The model of the CM is *red*, the NSM model is in *blue*, and the superposition of the two is in *black*. There are two ways to interpret this model, depending on whether the CM impact preceded or succeeded the NSM event

Figure 9.1 should be:

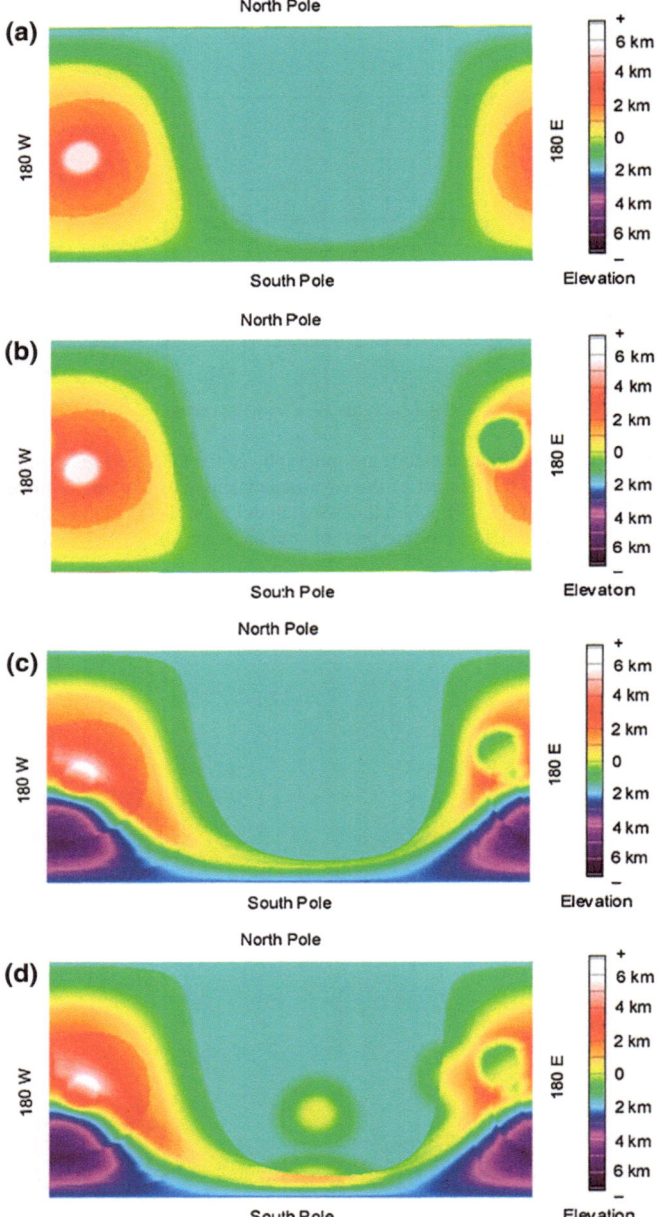

Fig. 9.1 The models of pre-Nectarisn age group 1, first NSM, then CM, then SPA and finally the four mounds. **a** Near side megabasin (*NSM*), **b** Chaplygin-Mandels'shtam Basin (*CM*), **c** South Pole-Aitken Basin (*SPA*), **d** Mounds

References

Andrews-Hanna, J. C., Zuber, M. T., & Banerdt, W. B. (2008). The Borealis basin and the origin of the martian crustal dichotomy. *Nature Leters, 453*, 1212–1215, 26 June 2008.

Andrews-Hanna, J. C. & Stewart, S. T. (2011). The crustal structure of Orientale and implications for basin formation. *LPSC2011*, Abstract # 2194.

Bottke, W. F., Vokrouhlicky, D., Minton, D., Morbidelli, A., Ramon, B., Simonson, B., & Levison, H. F. (2012a). An archaean heavy bombardment from a destabilized extension of the asteroid belt. *Nature, 485*, 78–81. doi: 10.1038. 2012, supplementary information, www.nature.com.

Bottke, W. F. (2012b). The great Archean bombardment. *LPSC 2012*, Abstract # 4036.

Byrne, C. J. (2004). Evidence for three basins beneath Oceanus procellarum. *LPSC 2004*, Abstract # 1103.

Byrne, C. J. (2005a). *Lunar orbiter photographic atlas of the near side of the Moon*. London: Springer.

Byrne, C. J. (2005b). Gravity focusing of swarms of potential impactors. *LPSC 2005*, Abstract # 1262.

Byrne, C. J. (2006). The near side megabasin of the Moon. *LPSC 2006*, Abstract 1930.

Byrne, C. J. (2007). A large basin on the near side of the Moon, Earth, Moon, and Planets, *101*, 153–188. doi: 10.1007/s11038-007-9225-8, 2007 (on line), 2008 (print).

Byrne, C. J. (2008). *The far side of the Moon: A photographic guide*. New York: Springer.

Byrne, C. J. (2012). Modeling the Moon's topographic features. *LPSC 2012*, Abstract # 1118.

Byrne, C. J. (2013). Evidence for Earth-accreting planetesimals intercepted by the Moon, LPSC 2013, Abstract #1344

Cadogan, P.H. (1974). Oldest and largest lunar basin? *Nature, 250*(5464), 315–316.

Compston, W., Williams, I. S., & Meyer, C. Jr. (1984). Age and chemistry of zircon from late-stage lunar differentiates. *Lunar and Planetary Science XI*, 182–183, LPI.

Cook, A. C., Spudis, P. D., Robinson, M. S., & Wattens, T. R. (2002). Lunar topography and basins mapped using a Clementine stereo digital elevation model. *LPSC 2002*, Abstract # 1281.

Crida, A. (2009). Solar System formation. *Reviews in Modern Astronomy,21*, 3008.

Croft, S. K. (1980). Cratering flow fields: Implications for the excavation and transient expansion stages of crater formation. *Proc. of the 11th Lunar and Planetary Science Conference*, Pergamon, 1980, pp. 2347–2376.

Frey, H. (2011). Previously unknown large impact basins on the Moon: Implications for lunar stratigraphy. In: W. Ambrose & D. Williams (Eds.) *Recent advances and current research issues in lunar stratigraphy*, GSA special paper 477, 2011.

Feldman, W. C., Gasnault, O., Maurice, S., Lawrence, D. J., Elphic, R. C., Lucey, P. G., & Binder, A. B. (2002). Global distribution of lunar composition: New results from Lunar Prospector. *Journal of Geophysical Research: Planets, 107*(E3), 5016. doi: 10.1029/2000JE001506.

Fernandez, V. A., Garrick-Bethel, I., Shuster, D. L., & Weiss, B. (2008). Common 4.2 Ga impact age in samples from Apollo 16 and 17. *LPSC 2008*, Abstract # 3028.

Garrick-Bethell, I. (2004). Ellipses of the South Pole-Aitken Basin: Implications for basin formation. *LPSC XXXV*, Abstract 1515.

Garvin, J. B., et al. (2011). Linné: Simple lunar mare crater geometry from LRO observations. *LPSC 2011*, Abstract # 2063.

Gomes, R., Levison, H. F., Tsigaris, K., & Morbidelli, A. (2005). Origin of the cataclysmic late heavy bombardment period of the terrestrial planets. *Nature, 435*, 4666–4669.

Grogan, M. (2006, August 21). Lunar Orbiter panel discussion, Museum of Flight, Seattle, Washington, talk and DVD.

Halliday, A. N. (2000). Terrestrial accretion rates and the origin of the Moon. *Earth and Planetary Science Letters, 176*(1), 17–30.

Hartmann, W. K., & Kuiper, G. P. (1962). Concentric structures surrounding lunar basins. University of Arizona, *Lunar and Planetary Laboratory Communications, I*(12), 51–66.

Hawke, B. R., Gguere, T. A., Gadddis, L. R., Smith, G. A. (2007). Remote sensing studies of Copernicus Rays: Implications for the Copernican–Eratosthenian Boundary. *LPSC 2007*, Abstract #1133.

Heiken, G. H., Vaniman, D., and French, B. M., eds. (1991). Lunar Sourcebook: A User's Guide to the Moon, CUP and LPI, 1991.

Hiesinger, H., & Head, J. W. III. (2006). New views of lunar geoscience: An introduction and overview, In: B. L. Jolliff, et al. *New views of the Moon, reviews in mineralogy and geochemistry* (Vol 60), mineralogical society of America.

Hikida, S. H., & Mizutani, S. (2005). Mass and moment of inertia constraints on the lunar crustal thickness: Relations between crustal density, mantle density, and the reference radius of the crust-mantle boundary. *Earth, Planets, and Space, 57*, 1121–1126.

Hikida, H. & Wieczorek, M.A. (2007). Crustal thickness of the Moon: New constraints from gravity inversion using polyhedral shape models. *LPSC 2007*, Abstract 1547.

Holsapple, K. A., & Schmidt, R. M. (1982). On the scaling of crater dimensions: 2. Impact processes, *Journal of Geophysical Research: Planets, 87*(B3), 1849–1870.

Housen, K. R., Schmidt, R. M., & Holsapple, K. A. (1983). Crater ejecta scaling laws: Fundamental forms based on dimensional analysis, *Journal of Geophysical Research: Planets, 88*(B3), 2485–2499.

Ivanov, B. A. (2007). Lunar impact basins–numerical modeling. *LPSC 2007*, Abstract 2003.

Kennedy, A. K. & de Laeter, J. R. (1994). The performance characteristics of the WA Shrimp II ion microprobe. *Proceedings of 8th International Conference on Geochronology*, cosmochronology and isotope geology, Berkeley, USA.

Lee, D-C., Halliday, A. N., Snyder, G., A., Taylor, L. A. (1997). Age and origin of the Moon. *Science, 278*, 1898–1103.

Lineweaver, C. H. & Norman, M. (2009). The Bombardment history of the Moon and the Origin of Life on Earth. In: *Australian Space Science Conference Series*: *8th Conference Proceedings NSSA*.

LROC (2012). LROC WMS image map. http://wms.lroc.asu.edu/lroc#damoon.

Losiak, A., et al. (2009). A new lunar impact crater database. *LPSC 2009*, Abstract 1532.

Maxwell, D. E. (1977). Simple Z model of cratering, ejection, and the overturned flap. In: D. J. Roddy et al. *Impact and Explosion Cratering*, 1003–1008, Pergomon.

Minton, D. A., et al. (2012). The early bombardment history of Mars revealed in ancient megabasins, Workshop on Early Solar System Impact Bombardment II, Abstract # 4040.

Muller, P. & Sjogren, W. (1968). Mascons: lunar mass concentrations. *Science, 161*(3842): 680–684.

Nakamura, R., et al. (2012). Compositional evidence for an impact origin of the Moon's Procellarum basin, Nature, Geoscience Letters, online, October 28, 2012.

National Space Science Data Center, NSSDC ID: 1966-027A (undated).

Nakamura, R., Yamamoto, S., Ishihara, Y., Yokota, Y., and Matsumaga, T. (2013). Differentiation of impact-generated magma seas on the Moon as revealed by spectral profiler onboard Kaguya, LPSC 2013, Abstract #1988.

Nemchin, A., et al. (2008). SIMS U-Pb study of zircon from Apollo 14 and 17 breccias: Implications for the evolution of lunar KREEP. *Geochimica et Cosmochimica Acta, 72*(2008), 668–689.

Nemchin, A., Timms, N., Pidgeon, R., Geisler, T., Reddy, S. and Meyer, C. (2009). Timing of crystallization of the lunar magma ocean constrained by the oldest zircon. *Nature Geoscience, 2*, 133–136.

Neumann, G. A., Zuber, M. T., Smith, D. E., & Lemoine, F. G. (1996). The lunar crust: Global structure and signature of major basins, *Journal of Geophysical Research: Planets,101*(E7), 16841–16843.

Neumann, G. A., Personal Communication, 2013.

Oberst, J., Unbekannt, H., Scholten, F., Haase, I, Hiesinger, H., Robinson, M. (2011). A search for degraded lunar basins using the LROC-WAC digital terrain model (GL.D100). *LPSC 2011*, Abstract # 1992.

Ohtake, M., et al. (2011). Vertical compositional trend within the lunar highlands crust. *LPSC 2011*, Abstract # 1169.

Petro, N. E. & Pieters, C. M. (2005). The lunar-wide effects of the formation of basins on the megaregolith. *LPSC 2005*, Abstract 1209.

Pidgeon, R. T. et al. (2010). Evidence for a Lunar "Cataclysm" at 4.34 Ga from Zircon U-Pb Systems. *LPSC 2010*, Abstract 1126.

Richardson, J. E. (2007). Improving the modeling of impact ejecta behavior: The effects of gravity and strength near the crater rim. *LPSC 2007*, Abstract #1345.

Sasaki, S., et. al. (2011). The first precise global gravity and topography of the Moon by Kaguya (Selene). JAXA, Kaguya archive web page: http://archive.ists.or.jp/upload_pdf/2011-k-04.pdf.

Schultz, P. H. (2007). Personal communication, September, 2007.

Sharpton, V. L. (1994). Evidence from Magellan for unexpectedly deep complex craters on Venus, In: B. O. Dressler et al. *Large meteorite impacts and planetary evolution* (pp. 19–27). Special paper 293, GSA.

Shearer, C. K. et al. (2006). Thermal and magmatic evolution of the Moon, in New Views of the Moon, ISBN 0-939950-72-3, *Mineralogical Society of America*.

Stewart, S. T. (2011). Impact basin formation: the mantle excavation paradox resolved. *LPSC 2011*, Abstract 1633.

Stewart, S. T. (2012a). Impact basin formation and structure from 3D simulations. *LPSC 2011*, Abstract 1633.

Stewart, S. T. (2012b). Dr. Sarah Stewart's web page, http://www.fas. harvard.edu/~planets/sstewart/Movies.html.

Stoffler, D., Ryder, G., Ivanov, B. A., Artemieva, N. A., Cintala, M. J., & Grieve, R. A. F. (2006) Cratering history and lunar chronology. In: *New views of the Moon*. ISBN 0-939950-72-3, *Mineralogical Society of America*.

Stuart-Alexander, D. E. (1978). Geologic map of the central far side of the moon, USGS Map I-1047, scale 1:5,000,000.

Taylor, G. J. (2006). Wandering gas giants and lunar bombardment. Planetary science research discoveries. http://www.psrd.hawaii.edu/Aug06/cataclysmDynamics.html.

Tera, F., Papanastassiou, D.A., Wasserburg, G.J. (1974). Isotopic evidence for a terminal lunar cataclysm. *Earth and Planetary Science Letters, 22*, 1–21.

Thompson, W. T. (1986). *Introduction to space dynamics*. New York: Dover.

Turtle, E. P., et al. (2000). Impact structures: what does crater diameter mean? LPSC 2004.

Tsiganis, K., Bomes, R., Morbidelli, A., & Levison, H. F. (2005). Origin of the orbital architecture of the giant planets of the Solar System. *Nature, 435*, 459461.

Vaughan, W. M., Head, J. W., Hess, P. C., Wilson, L., Neumann, G. A., Smith, D. E., & Zuber, M. T. (2012). Depth and differentiation of the Orientale Melt Lake. *LPSC 2012*, Abstract # 1302.

Whitaker, E. A. (1981). The lunar Procellarum basin, in multi-ring basins. *LPSCP 12*, Part A.

Wieczorek, M. A., & Phillips, R. J. (1998). The structure of lunar basins: Implications for basin forming processes. *LPSC XXIX*, Abstract 1299.

Wieczorek, M. A., Phillips, R. J., Korotev, R. L., Jolliff, B. L., & Haskin, L. A. (1999). Geophysical evidence for the existence of the lunar Procellarum KREEP terrane. *LPSC XXX*, Abstract 1548.

Wieczorek, M. A., & Phillips, R. J. (2000). The Procellarum KREEP terrane; implications for mare volcanism and lunar evolution. *Journal of Geophysical Research: Planets,105* (E8), 20417–20430.

Wieczorek, M. A. et al. (2012). The crust of the Moon as seen by GRAIL, Sciencexpress Reports, December 5, 2012, pp. 1–10.

Wilhelms, D. E., Howard, K. A., & Wilshire, H. G. (1979). Geologic map of the south side of the Moon, USGS Map I-1162 scale 1:5,000,000.

Wilhelms, D. E., & Squyres, S. W. (1984). The martian hemispheric dichotomy may be due to a giant impact. *Nature309*, 138–140.

Wilhelms, D. E. (1987). The geologic history of the Moon, USGS professional paper 1348, US Gov. Printing Office.

Wilhelms, D. E. (2012). Personal communication, August, 2012.

Wood, C. A. (2003). Observing the Sky (blog), noted in Lunar Basins List, http://the-moon.wikispaces.com/Lunar+Basins+List.

Wood, J. A. (1973). Bombardment as a cause of the lunar asymmetry. *The Moon* 8, LPI.

Zuber, M. T., Smith, D. E., Neumann, G. A. (1994). The shape and internal structure of the Moon from the Clementine Mission. *Science, 266,* 1839–1843.

Index

C. J. Byrne, *The Moon's Near Side Megabasin and Far Side Bulge*,
SpringerBriefs in Astronomy, DOI: 10.1007/978-1-4614-6949-0,
© Charles J. Byrne 2013